HUMBUG

MG

上汽名爵

MG CENTURY

100 YEARS—SAFETY FAST!

DAVID KNOWLES

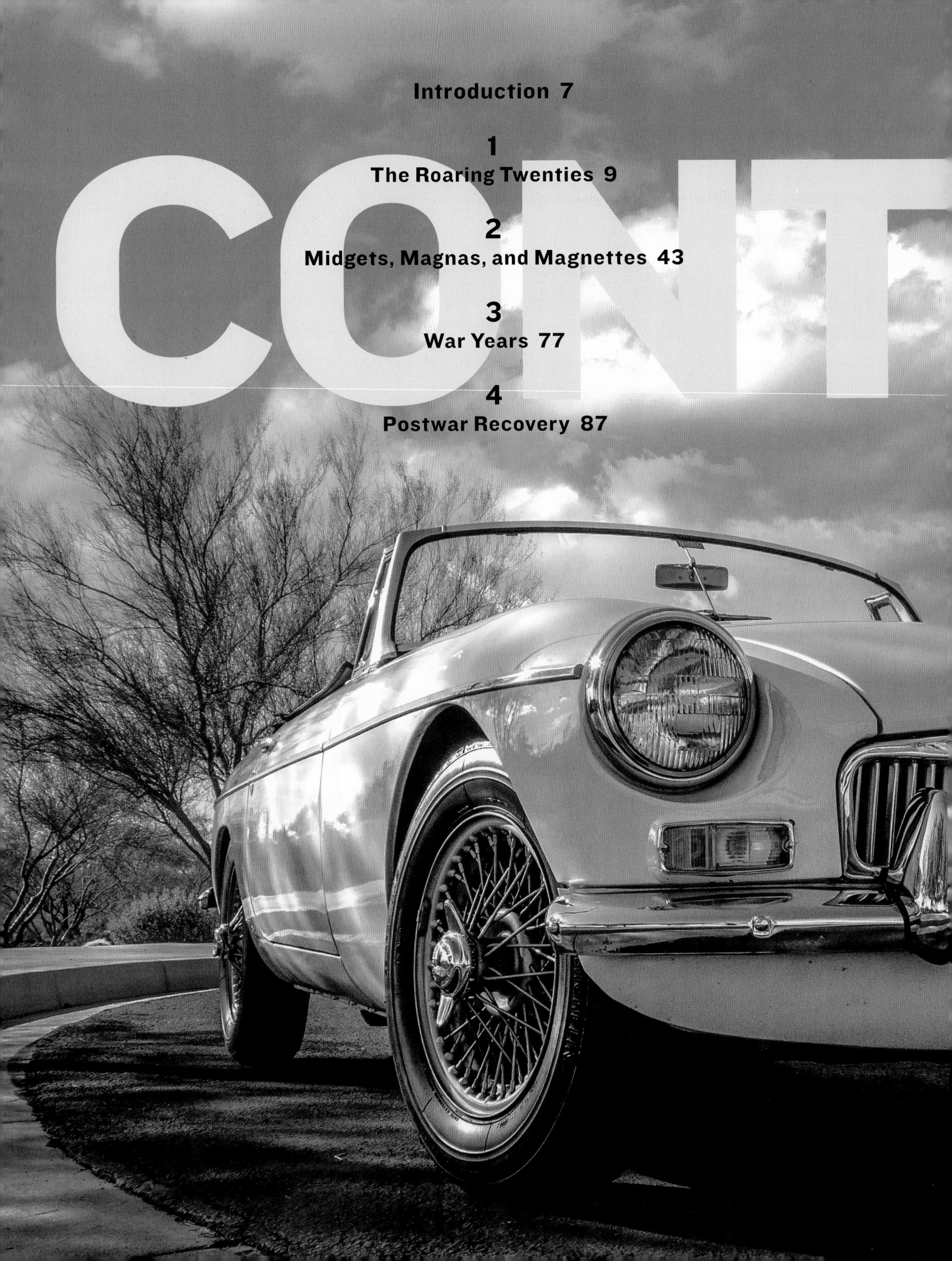

CONT

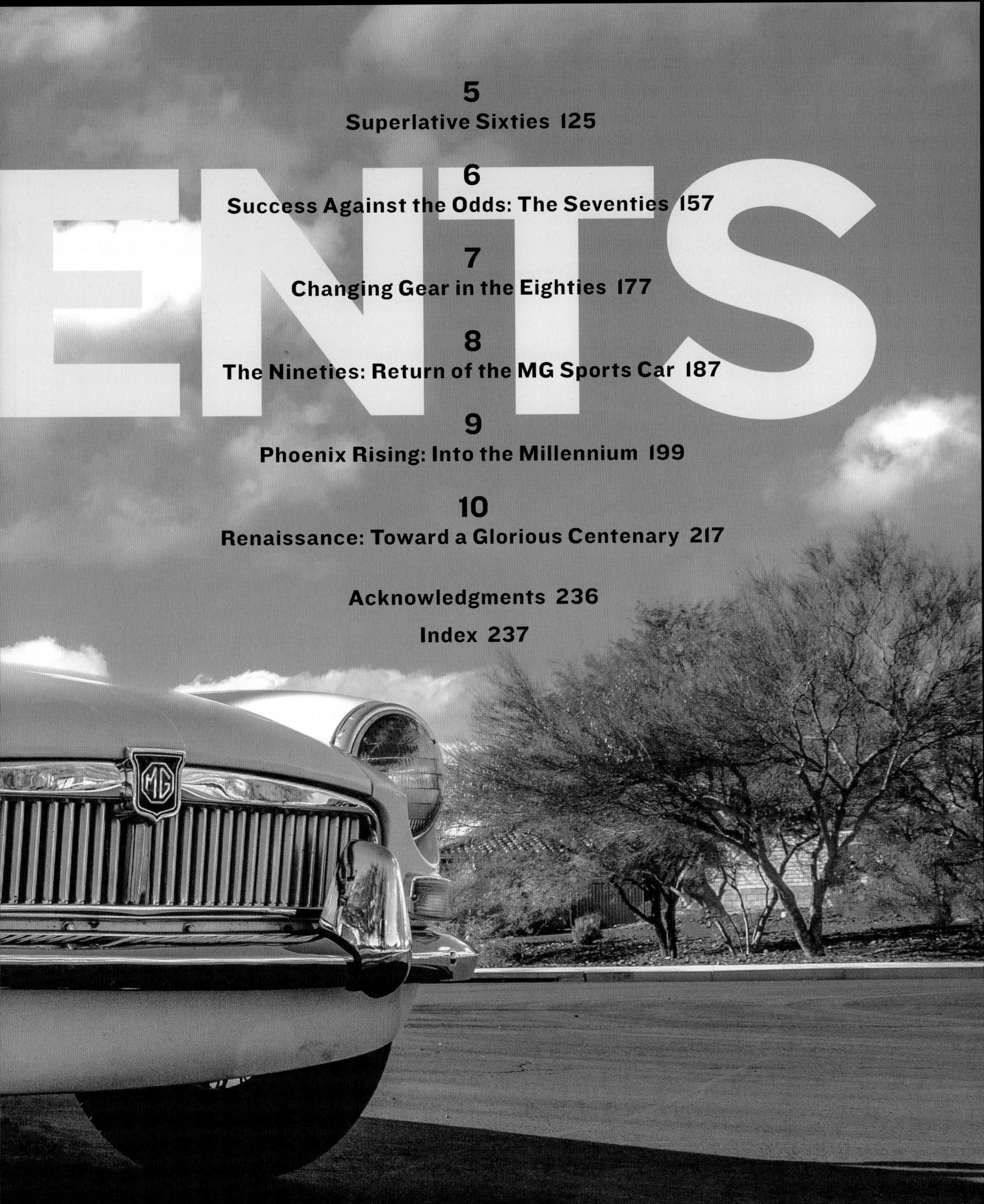

ENTS

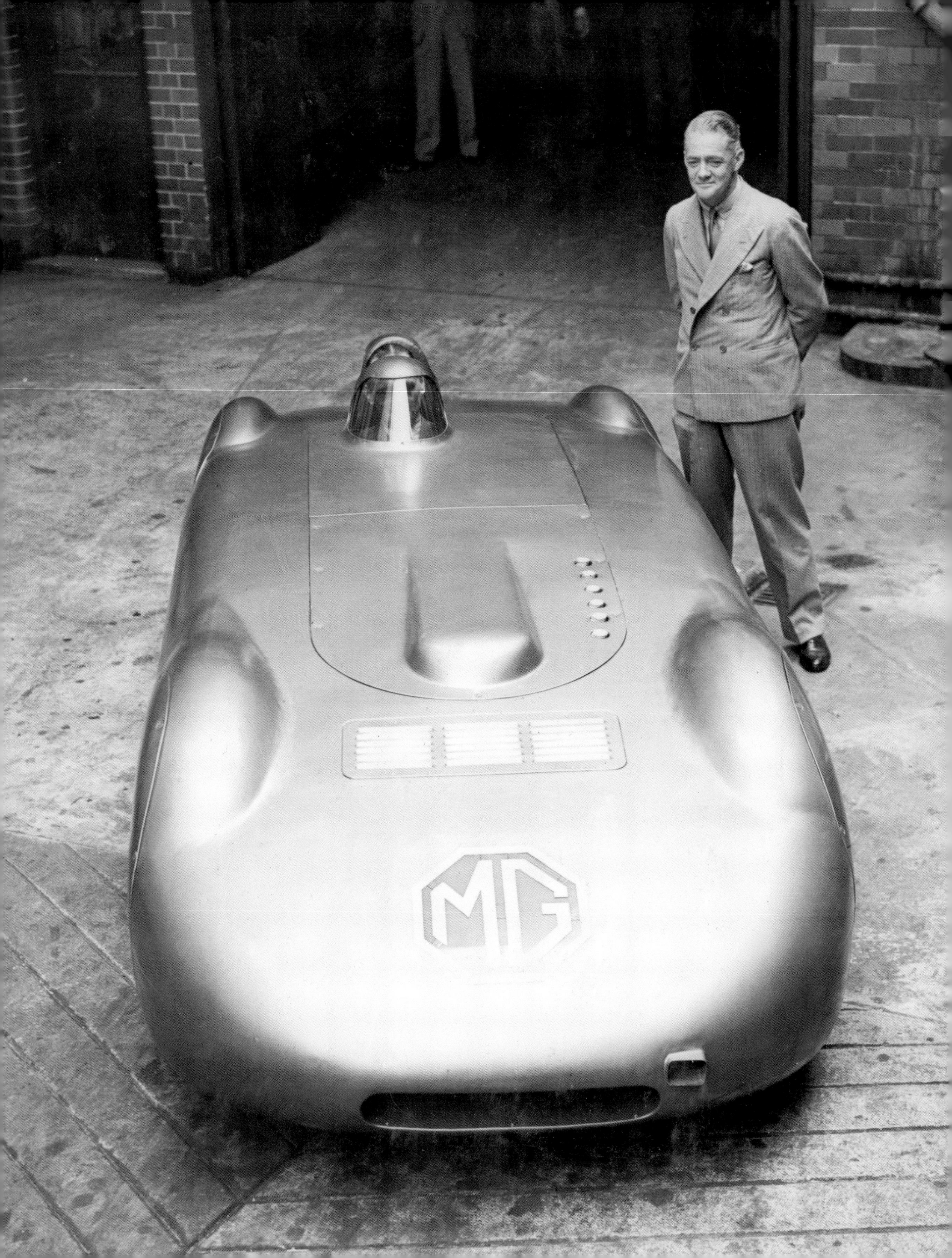
MG

Introduction

In so many ways, the world of one hundred years ago was a very different place, but cultural aspirations were fundamentally less different than we may think. Exciting innovations, the motorcar among them, were welcome by people keen to embrace what was new. These man-made wonders would evolve and occasionally lead to a kind of innate tribalism as brand loyalties took hold. Such passions have always crossed many fields, but the emergence of practical and affordable motorcars in the early part of the twentieth century led to an explosive growth of business, accompanied by customer enthusiasm and support.

William Morris, who began his career repairing and making bicycles, was thirty-five by the time he made his first motorcar and well knew the importance of cultivating a sense of marque loyalty by offering the best combination of value and quality to those who bought his products. It took another man, by the name of Cecil Kimber, to recognize how visual flair and sportiness added desire to the underlying basic qualities of Morris's offerings. It was a wonderful cocktail of talents while it lasted, and in those heydays of the late 1920s and 1930s "MG" become a watchword for sports, racing, and record-breaking cars. The personal and financial backing underlying the patronage helped ensure that MG came to represent some of the finest sporting cars of the time.

Like many such stories, the tale would not continue uninterrupted, and there have been many twists and turns over the past century. Despite it all, the flame that Cecil Kimber kindled burned brightly long after he had gone (forced out mostly through jealousy). Even lacking Kimber at the helm, the MG passion continued to thrive and spread to generations of owners and enthusiasts around the world. Post WWII, it was Americans in particular who discovered MG and took the small sports cars to heart. Still, the run of ever-growing success was eventually broken by external forces, and MG sales ended in North America over forty years ago.

Even the cataclysmic closure of the famous Abingdon MG factory in 1980 was not the end of the story, and the marque continued to thrive over subsequent decades, even as many other famous makes departed the scene. There have been occasional gaps in MG's history when sports cars have been absent, though thankfully they are back again as an integral part of the modern range.

Nowadays, the MG marque is making new friends in China, India, Thailand, and other territories that seldom knew the marque previously. What is especially remarkable—and poignant too, mindful of what else has been lost—is the fact that what began in the mid-1920s as an offshoot of William Morris's main business enterprise has evolved to the point where it is not only one of the few names from the historic line that has lasted the course, but has become the proud export pinnacle of one of the world's largest carmakers, SAIC Motor of Shanghai. It is a remarkable story, with many twists and turns, and I hope you will find this book to be a worthy guide.

—David Knowles, Ruislip, U.K., 2023

OPPOSITE: Lord Nuffield stands to one side at the press luncheon at University Motors in June 1948, which saw the unveiling of the exciting new record car for Goldie Gardner (seated in the cockpit). The first serious outing for this vehicle would be at Frankfurt in September. *Enever Family Archive*

BP
THE SPORTING "SPIRIT"
START
BP
38
FC 4442

1

The Roaring Twenties

MORRIS ORIGINS

The first part of the story of Morris Garages, which begat MG, begins with the origins and early life of the company's founder himself, William Richard Morris. Much has been written about Morris, whose singular vision and pugnacious determination sowed the seeds for MG, a car manufacturer that continues to thrive in ways that its founder could never have imagined.

The fact that MG has outlived his own Morris Motors would undoubtedly amaze him (as much as its foreign ownership would disappoint him), but it is also a great shame that his works are barely celebrated in the city where he started his business. Many ignorant of his life's work and charitable beneficence may even confuse him with his earlier—and admittedly equally influential—namesake, William Morris, a leading figure in the Arts and Crafts Movement of the nineteenth century.

When he became a baronet, he took the name of the small Oxfordshire village of Nuffield, where he and his wife had made their home. Those who have been treated at one of the Nuffield hospitals, benefited from a Nuffield scholarship, or studied at the Oxford University college founded with his support that also bears his name, may be unaware of these institutions' connection with the British car industry, including the products described in this book.

His family traced its origins in the Oxford area, although Morris was actually born in a small house in the small village of Hallow, northwest of Worcester, in 1877.

OPPOSITE: We tend to associate MG and Morris Garages as being the main outlet for early motorsports. Morris Oxfords and Cowleys were unaffiliated with Morris Garages, but they also competed in early races, as shown here. *Enever Family Archive*

William R. Morris on an early bicycle, ca. 1895. This image first appeared in public in the 1930s along with speculation that Lord Nuffield might have been contemplating returning to bicycle manufacture. *Author archive*

The family soon moved back to Oxford and Morris grew up in the area, appearing in the 1891 census as still at school and living with his parents and sisters in a cottage in Brasenose Lane. His father had taken up a post as the town's farm bailiff. At the age of fifteen, Morris left school to help the family's finances and took an apprenticeship with a bicycle maker in St Giles' Street. A year later, he left the cycle shop after being refused a raise, deciding henceforth to always work for himself. From the 1901 census we learn that Morris (now twenty-three years old), described as a "cycle manufacturer and employer," was living with his father Frederick (retired from farming and now described as a cycle agent), mother Emily, and sisters Alice and Emily, all living at 16 James Street in East Oxford.

A keen cyclist who enjoyed building and racing (with distinction) his own bicycles, Morris joined Joseph Cooper in 1901 to form Morris & Cooper, located at 48 High Street in Oxford. The arrangement was not destined to last: Morris wanted to build motorcycles, but Cooper did not relish the risks involved in creating the newly motorized vehicles. Before long, Morris bought out Cooper, although they remained on good terms. Cooper would later work at MG's Cowley plant, serving there for the remainder of his career. This set a pattern for consistently rewarding his loyal partners and employees, something Morris did throughout his career.

In 1903, Morris joined an established cycle retailer, Frank Barton, and an Oxford undergraduate, Walter Launcelot Creyke (their silent partner). The trio formed the Oxford Automobile & Cycle Agency to meet Morris's goals of expanding his motorcycle offerings and branching out into cars. Unfortunately, Creyke was more interested in entertaining his friends and associates without much thought to actual outlay. The business failed after a year and Creyke went to Manitoba.

Miles Thomas, who worked for Morris from 1924 to 1947, recalled in his autobiography, *Out on a Wing* (published in 1964), what happened when the business failed. "Young Morris had the mortification of having to stand in the rain at an auction sale of the assets of the partnership; he had to borrow £50 to buy back his own kit of tools, which to him were his stock-in-trade. With this £50 debt hung round his neck like an albatross, he had to start again from the bottom rung of the ladder."

Dusting himself off, Morris formed a new partnership with George Cooke, a wealthy tobacconist and businessman who ran a nearby shop. With their agreement signed on June 8, 1905, Morris managed to rebuild his cycle shop interests and was soon firmly on the way back to prosperity. This partnership too was dissolved, albeit amicably, less than two years later on April 20, 1907. Meanwhile, Frank Barton, as we shall see, continued to play a part in the Morris story.

In the same year as the failure of his partnership with Cooke, Morris managed to buy up a group of tumbledown sheds in Longwall, just around the corner from High Street, and opposite the walls of Magdalen College. He soon had the sheds rebuilt and transformed into the fledgling Morris Garage. His timing couldn't have been better, and business boomed: He offered a space and services that combined the popularity of newly available motorcars, the proximity of wealthy undergraduates (who needed somewhere to work on, garage, or maintain their motorcars), and the demand for specialists who could fulfill all these needs.

MORRIS ON FOUR WHEELS

Selling and servicing other manufacturers' cars brought William Morris success—and the business model would serve him well for many years—but having built his own bicycles and motorcycles, he was determined to build his own motorcar. And what better name for it than his own? In October 1912, Morris formed WRM Motors Limited and, at the London Motor Show at Olympia in November of that year, he enticed a number of early backers to place orders for his new Morris Oxford car on the basis of blueprints alone. The resulting Bullnose Morris first appeared the following March, priced at £175, complete with its distinctive rounded radiator made of polished "German silver" (actually an alloy of nickel, copper, and zinc).

The bodies of these Bullnose Oxfords came from Hollick & Pratt of Coventry, as well as some from Raworths, the Oxford-based coachbuilders. The cars' 10-horsepower engines, carburetors, and three-speed gearboxes were specially designed to Morris's requirements and manufactured by White & Poppe in Coventry; Morris paid them £50 per car—almost 30 percent of their overall price.

That same year, the business was renamed Morris Garages—Proprietor W.R. Morris. The headquarters, located on Queen Street, included a car showroom and a downstairs salesroom for Sunbeam motorcycles. Later, in 1914, Morris took over a yard behind the Clarendon Hotel and turned it into a workshop.

In February 1914, Morris visited the United States for the first time to meet potential automobile component suppliers. The Continental Motor Manufacturing Company offered him a small-capacity engine (1,495 cc), adding that they could supply the engine along with a gearbox from the Detroit Gear Machine Company for about half the cost of the White & Poppe units. Saving £35 was certainly attractive, even allowing for shipping costs, so Morris negotiated a supply contract.

The Continental engine was reserved for a new, less expensive Morris, the Morris Cowley, which would come to be seen as WRM's masterpiece. Featuring a Hollick & Pratt body, the car was announced in 1915. As motoring historian Peter Seymour points out, this Morris Cowley and its Oxford derivative formed the basis of the first true MGs—though that's getting ahead of the story.

Members of the Morris Garages Longwall Garage staff on the occasion of the first works outing to Brighton in 1910. *Left to right:* Walter Abbott, Bertram "Copper" Crease, E. J. Tobin, R. A. Bishop, Joe Cooper, and William R. Morris. *Enever Family Archive*

By this time, war in Europe had broken out, and its effects were soon felt throughout England, including in the budding automobile industry. In September 1915, British Chancellor of the Exchequer Reginald McKenna imposed a swingeing one-third import tax on American goods, followed the next spring by a government ban on shipments of nonessential supplies.

Morris had reason to be thankful that he was busy with war work. To circumvent the punitive taxes, he sought someone to build the Continental engines in the U.K. under license. The logical partner for Morris was the French company Hotchkiss et Cie., which had set up in Coventry in 1915 to offset the risk of its Paris machine-gun factory's potential capture by the Germans. Even with this relationship, it was not until after the Armistice of 1918 that the Morris Cowley began to be seen in any numbers. The first Hotchkiss engines were delivered in July 1919, now that the ready market for Hotchkiss's machine guns had dried up.

A month before the engines appeared, Morris had decided to place W.R.M. Motors into voluntary liquidation and immediately open a new business, Morris Motors Limited. The underlying reason was to disengage himself from what he saw as restrictive obligations to his original distributors and reach out to new manufacturers offering the latest technology.

It was a situation fueled by the recent world war's disruptions and the vanishing of the old world order, itself largely formed in the Victorian era and the brief Edwardian decade. Early-twentieth-century Europe in particular was reeling from the carnage of war, followed by the catastrophic 1918 influenza. These extensive disasters had decimated national economies and families alike. The interlude of global conflict had nevertheless seen a dramatic acceleration in and the availability of new technology. The Great War had led to remarkable progress in motorized transportation on land, sea, and air. At the heart of this pulse of change was the rise of the automobile, or motorcar as the British preferred to call it.

The original Longwall garage, ca. 1911. *Author archive*

MORRIS THE ENTREPRENEUR

With technical revolutions there tend to be *innovators* and *entrepreneurs*. Innovators formulate new ideas and devise technology in general, while the entrepreneurs ride on their inventive counterparts' coattails, spotting and quickly exploiting a sound commercial opportunity. If Austin Motor Company founder Herbert Austin could be considered an innovator, William Richard Morris was at heart an entrepreneur.

No more a classically trained engineer than most of his peers, Morris gained mastery of a field, then spotted opportunities for improvements, and finally exploited his vision to create a revolutionary product. Starting with the bicycle, which he had raced with some success, Morris went on to servicing and selling examples made by others, then assembling and selling his own. When he saw still greater opportunities in the new motorcycles, then the even more promising four-wheel motorcar, Morris took advantage of his position within the evolving motorized vehicle industry to meet the growth in small car sales.

The Morris Cowley's appearance in 1912 marked his entry into that market, but production of that car was overshadowed by looming war clouds. Still, Morris's business model proved strong enough to weather the storm: Whereas some of his rivals—notably Austin—preferred to make most of the key parts for what they actually sold, Morris was by habit an assembler, buying the best and most affordable components and securing the best prices in the process. Undoubtedly this stemmed in part from his bicycle business, but it was aided in no small part by his farsighted selection of suppliers to buy out and restructure to his own specifications, often retaining the original staff in hands-on managerial control. In this way Morris built an empire largely by accretion.

This in turn meant that Morris presided over a complex mixture of businesses, some absorbed into the core Morris Motors enterprise, several others owned personally by Morris himself. It meant that, to an extent, he was able to dip into (and interfere with) his various enterprises, which were often staffed by his acolytes. One such business was the Morris Garages, which had grown by osmosis from his original retail and servicing business. Soon the work was spread across a number of sites in the center of Oxford, cheek by jowl with the gleaming spires of academia that housed many of his early clientele.

It was not all plain sailing, however: The start of the first new decade after World War I saw recession and financial strife. Depression shook world economies beginning in 1920, running from 1920 to 1921 in the United States and for a year longer in the United Kingdom, the latter having experienced a short-lived economic boom in 1919–1920. Morris was affected in particular, as demand for all-new cars plummeted in late 1920 and nearly precipitated both Austin and Morris into the hands of the receivers. Morris survived largely because he was able to secure generous bank guarantees, slash retail prices, obtain credit extensions from his suppliers, and convince his distributors to accept a lower profit margin.

Further help came with a rebound that took place immediately after the economic downturns of the early 1920s. A newly emboldened young generation was ready to seek out novel attractions. They embraced with youthful vigor modern dance, music, radio, moving pictures, and more affordable motorcycles and small automobiles, the latter clearly aimed at less affluent middle-class owner-drivers rather than the upper crust and their chauffeurs.

CECIL KIMBER ARRIVES

In 1921 Morris brought on Cecil Kimber as his new sales manager. Kimber came from E. G. Wrigley & Co., an automobile manufacturer that had overstretched itself and, in the process, taken most of Kimber's savings with it. After descending further into financial difficulties,

ABOVE: A fleet of hire cars outside the later, much grander, Longwall premises of Morris Garages. Although subsequently repurposed, the buildings largely survive to this day. *Author archive*

BELOW: A 1912 Morris Bullnose Oxford. *Author archive*

Cecil Kimber at his desk in what later became known as Larkhill House, the headquarters of the Abingdon MG factory. The mascot (at right, slightly out of focus) is believed to be a *panthère sautant* ("leaping panther") by French sculptor Casimir Brau, which predates the more familiar Jaguar mascot adopted by William Lyons. *Author archive*

Wrigley was put up for sale in 1923 and Morris bought the company the following year.

There had been a few management shake-ups at Morris Garages in the years leading up to Kimber's appointment, and more instability was to come before the company's leadership would settle into place. Frank Barton, Morris's former business partner, was the first to leave.

A small advertisement in the *Oxford Times* of January 18, 1919, stated that "Mr. W. R. Morris wishes to inform his many customers and friends that Mr. E. C. Armstead, who is well known to many of the citizens of Oxford and district, has been appointed 'General Manager of the Morris Garages' in the place of Mr. F. G. Barton, who owing to ill-health was forced to relinquish the position." Interestingly, Frank Barton remained in Morris's circle, going on to found one of the first Morris Motors appointed agents for Morris cars in Plymouth, Devon.

Edwin Clipsham Armstead, himself a former manager of his own cycle shops in Oxford and a member of the Oxford City Council's West Ward, had done business with Morris before taking his post at Morris Garages. In early 1921, though, Armstead appears to have stepped back from the general manager role, which led Morris to give Carl Breeden overall control of Morris Garages. Until recently the sales director of Joseph Lucas before a falling out with his brother-in-law, Oliver Lucas, Breeden was one of a handful of businessmen who had extended a generous credit line when Morris had been in trouble in 1919. William Morris seldom forgot such favors, and Breeden remained a lifelong friend.

Around this time, Albert Sydney "Syd" Enever arrived at Morris Garages. A keen lad of fourteen, Enever's relationship with Morris Garages would be critical for much of the MG story. This was a year that also saw the Morris Garages growing, with premises already established in the High Street, Longwall, and the yard behind the Clarendon Hotel in Cornmarket.

Carl Breeden's tenure as general manager was short-lived. He evidently saw more attractive opportunities back home in Birmingham, where he bought into a business that would become the highly successful Wilmot-Breeden auto component manufacturing firm. Much of Morris's business went to Wilmot-Breeden's way over the following years.

With Breeden gone, Kimber lobbied Morris for the lead post, and in March 1922 he found himself appointed general manager. Armstead also left that month and not long after committed suicide. According to the inquest into his death, he had regretted his business decisions before his move to Morris Garages: "[H]e worried a lot about giving up his business. He thought he had made a great mistake. He gave up his business in Broad Street during the war, and just recently he had been somewhat depressed." The flag at Oxford's Carfax Tower was flown at half-mast—a sure mark of the respect and esteem in which Armstead was held.

Kimber wasted no time putting his stamp on Morris Garages' business following the departure of Breeden and Armstead, working closely with Morris when the latter was not busying himself at his Cowley works. From the outset, Kimber was keen to find a way to boost sales of the cars that bore his patron's name. Not unlike many garage proprietors before and after, he selected fancy paint and trim, more rakish bodywork, and tuned components designed to improve the basic Morris vehicle. In the process, his new styling selections allowed the company to charge the customer for a special, superficially bespoke motorcar. These Morris Garages Specials soon formed a niche market, selling mostly in the Oxford area. With their bright paint colors and polished body panels and wheel trim covers, these models caught the eye of moderately affluent customers, among them university undergraduates.

A fashionable body type that emerged at the time was known as the "Chummy," a nickname derived from the cozy nature of its effectively two-seat interior, with all-encompassing hood and occasional rear seating. A letter to the *Autocar* of October 27, 1922, signed "The Morris Garages" (we may safely assume the author was Kimber) heralded the arrival at the forthcoming Motor Show (and, by implication, at Morris Garages Oxford showrooms) of "that type of body which is now known, for want of a better name, as the 'occasional four,' in which a small space covered by the hood behind the front two seats provides accommodation for two or three children, for one adult, or luggage."

In the *Oxford Times* of December 8, 1922, an advertisement from Morris Garages appeared offering the chance to "have a 'chummy' body on your Morris Cowley." Its price was quoted as "255 Guineas complete on a 1923 Morris Cowley four-seater chassis," which happened to be the same price as the standard Morris Cowley four-seater tourer. It is important to note that this was no "MG" as such: That name and the badge forever associated with it had yet to appear. As for the Chummy bodies themselves, it seems that they were not exclusive to Morris Garages, being marketed as well by Parkside Garages of Coventry. Regardless of this detail, it was Morris Garages that sold them with moderate success in the Oxford area.

This slightly more personal version was what Kimber first saw as the basis for a sporting conversion of the Morris Cowley, rather than the four-seater Oxford. From William Morris's perspective, if this added conquest-sales at no cost to his core business, he would be content.

THE ORIGINS OF THE MG BADGE

Kimber embarked on a subtle campaign to build the identity of Morris Garages within months of taking his new position. His focus was on creating what we would now call a brand, one aspect of which was the introduction of a logo in the form of an octagon with the letters *M* and *G* fitted inside. Various authors of this simple, clever piece of art deco design have been named, but the general consensus is that credit probably belongs to Ted Lee, who staked his claim to author Jonathan Wood in 1982.

At the time the logo appeared, Lee was an accountant at Morris Garages. As he told Wood, "I drew out this badge with a little ruler I'd bought from high school—I was good at art . . . Kimber saw it and said 'that's just the thing.'" Next, the badge was shown to Morris, who apparently said it was "the best thing to come out of the company . . . and it will never go out of it." For many years, it was often claimed—erroneously, as we shall see—that the badge's debut before the general public was in May 1924, when it appeared in a Morris Garages advertisement in the *Morris Owner*.

It so happens that this was the date subsequently referenced when the new logo was formally registered nearly four years later, on April 2, 1928. The official "Class 22" entry at the British Patent Office proclaimed that "User claimed from May 1st 1924. . . . The applicants undertake that this Mark, when registered, and the Mark No. 490,090, shall be assigned or transmitted only as a whole and not separately." From this, one may deduce that the arrangement was dependent upon the MG badge being used whole and intact, with the letters within a frame and not without it: This

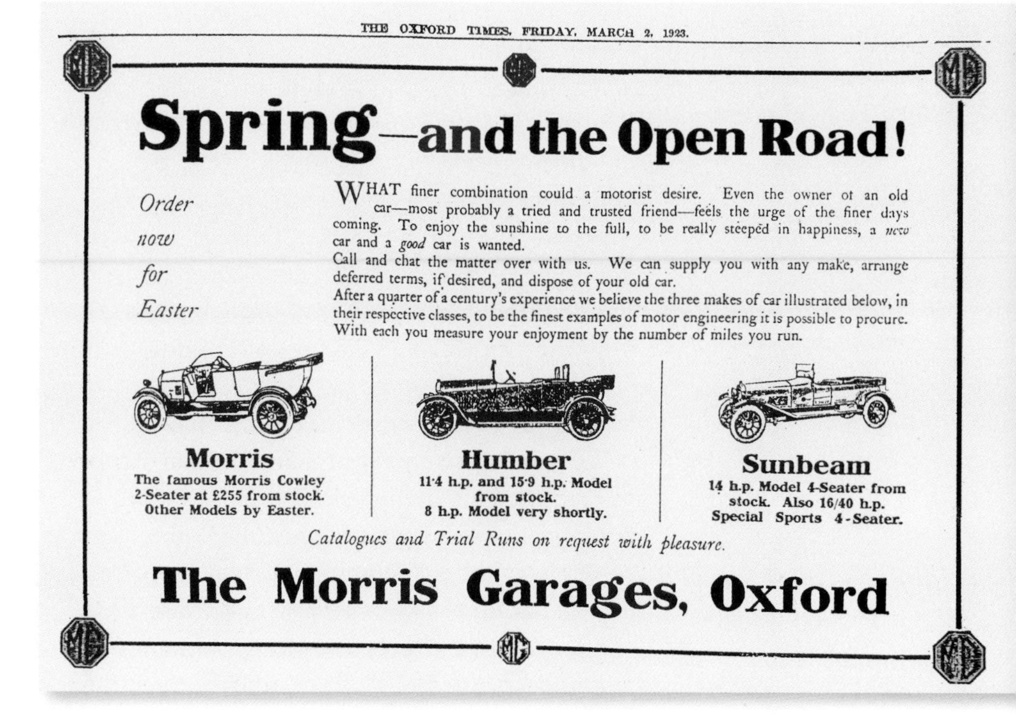

Believed to be the first Morris Garages advertisement to feature the octagonal MG logo, from the *Oxford Times* of March 2, 1923. Note that the ad makes no mention of any bespoke MG cars. *Author archive*

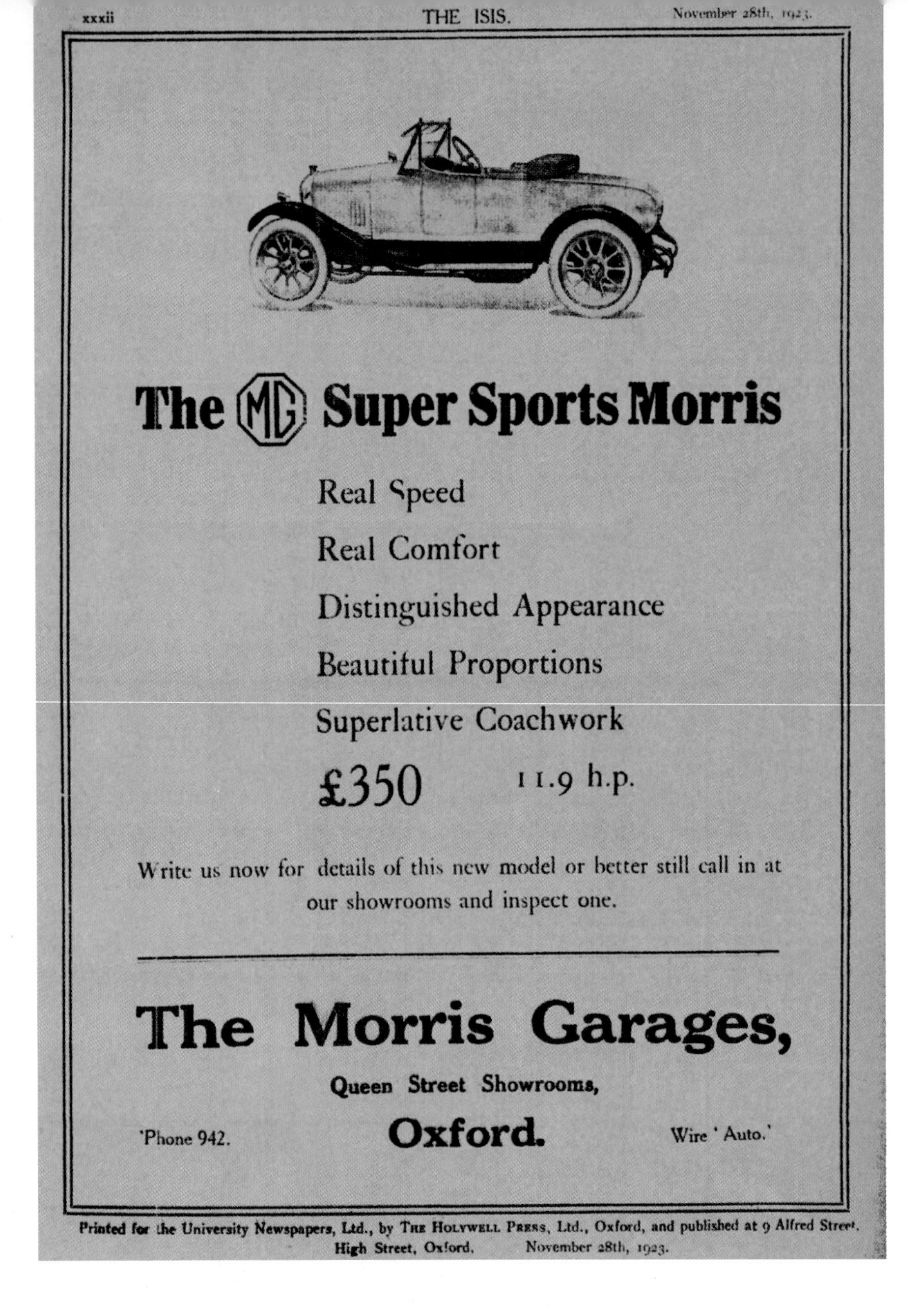

xxxii THE ISIS. November 28th, 1923.

The MG Super Sports Morris

Real Speed
Real Comfort
Distinguished Appearance
Beautiful Proportions
Superlative Coachwork
£350 11.9 h.p.

Write us now for details of this new model or better still call in at our showrooms and inspect one.

The Morris Garages,
Queen Street Showrooms,
'Phone 942. Oxford. Wire 'Auto.'

Printed for the University Newspapers, Ltd., by The Holywell Press, Ltd., Oxford, and published at 9 Alfred Street, High Street, Oxford, November 28th, 1923.

ABOVE: An ad on the back cover of the November 28, 1923, issue of the *Isis*. Even at this stage, the MG badge was portrayed as a part of the name of the car: the MG Super Sports Morris. Cecil Kimber eventually went so far as to acquire a typewriter with a special "MG" octagon key. *Author archive*

RIGHT: On the way toward the true MG identity—but not quite. This badge is from a 1924 MG Super Sports Morris Oxford. As shown in chapter 8, the Super Sports name briefly reappeared. *Author archive*

would prove to be an important stipulation that resonates even a century later.

Notwithstanding the popular currency generated by this document, apparently confirming May 1924 as the birth date of the badge, much earlier uses of the MG octagon have since been discovered. Some early claims attest to its debut on the previous November, but evidence found by authors Robin Barraclough and Phill Jennings eventually proved that its origins go back even earlier. The March 2, 1923, edition of the *Oxford Times* includes a print appearance in the form of an advertisement that displays a version of the logo. The ad carries the message "Spring—and the Open Road!" with small line drawings representing the Morris Cowley, Humber (offered as 11.4- and 15.9-horsepower models), and a four-seater sporting 14-horsepower Sunbeam. The advertiser requests the would-be customer to "order now for Easter." Importantly, however, the ad's rectangular frame features six of the new octagonal MG badges dotted around the edges, one at each corner and slightly smaller ones at the center points of the top and bottom sides of the frame itself.

A few weeks later, a further advertisement, in the April 13, 1923, edition of the *Oxford Times*, presents a simple celebration of the output of the Morris Garages patron's main business, describing Morris cars in general: "without the slightest exaggeration the Morris Car for 1923 is the most outstanding value ever offered to a car-buying public." There are no references to other makes in the ad. Again, the border of the advertisement features six MG octagons, in this instance, all of the same size.

By the time their advertisement in the *Oxford Times* of May 4, 1923, appeared, Morris Garages had added Dodge to the previous three marques referenced two months earlier, boasting that the practice by the business of "concentration on the sales and service of just four makes of Motorcar has undoubtedly made us the largest automobile distributors in the three counties" and adding that "we know that nothing finer is made in their respective classes." Once more, the advertisement was framed with six octagonal MG badges—still a year before the date claimed in the patent application.

The same logo was still being used in other advertisements placed in the Oxford students' magazine the *Isis* of November 21, 1923, and the subsequent one of December 5. In these, however, the MG badge was used as a graphic feature within the heart of the

advertisement itself, the badge replacing the usual Morris Garages text in the description of the car being advertised, rendering it boldly and unambiguously as "The 'MG' Super Sports Morris." This advertisement, taking up a full page in the magazine, describes these cars as possessing "real speed, real comfort, distinguished appearance, beautiful proportions and superlative coachwork," while customers are directed to the Queen Street Showrooms in Oxford.

Turning back to the May 1924 advertisement in the *Morris Owner*, we are left to wonder how the logo's appearance in that issue became the first instance of the octagonal badge in Morris Garages' collective memory. This internally funded magazine was edited by Miles Thomas, who had been lured by William Morris from *Motor Magazine* with a salary that equaled £87,000 (around $350,000) in 2023 currency.

The first issue of *Morris Owner* was published in January 1924, with "The MG Super Sports Morris" as the featured car. This vehicle, one of the first Raworth-bodied Cowleys, offered a 11.9-horsepower car that could "climb the famous Porlock Hill at 25 miles per hour." (The issue's cover story helpfully pointed out that this steep hill in bucolic English Somerset had a slope of one in five, a 20 percent grade.)

"Mounted on this out-of-the-ordinary chassis is the most delightful two-seater body imaginable," the text gushed. It's likely that these cars were the first to actually feature the octagonal MG badge, though only on the body plate rather than in a more prominent location. The paean of praise continues: "Beautifully comfortable, with adjustable seat and single dickey, the finish is of the highest class and the style irreproachable. The 'tout ensemble' is one of the finest productions we have ever turned out from our famous Queen Street Showrooms. For a car of such distinction the price, £350 is extraordinarily modest. May we send you further particulars?"

It seems reasonable to suppose that when Kimber and Morris were registering the new badge in 1928, they simply remembered its appearance in the May 1924 *Morris Owner* (or were better able to evidence this) and had either forgotten the earlier 1923 instances or, equally plausibly, simply had no copies of the requisite publications on hand to reference.

BUILDING THE MORRIS GARAGES SPECIALS

Building the Specials required expertise that was slightly above that of the usual service mechanic, and as well as skill the work

LEFT: The final steps toward the familiar octagonal MG badge: The enamel badge has replaced the earlier blue Morris Oxford hybrid. On this car, the MG letters also appear on the face of the radiator as cut-out letters in what was known as "German silver," a nickel alloy. *Author archive*

This MG Super Sports Morris
will climb the famous Porlock Hill
at 25 miles per hour

THE gradient of this noted acclivity is one in five and the Treasury Rating of the car is only 11.9 h.p. It will be seen therefore that the inherent possibilities of the famous Morris engine can be brought out by those who know how

The result is an exceptionally fast touring motor car capable of 60 miles an hour on the flat, and wonderful acceleration. The modified steering and springing gives a glued-to-the-road effect producing finger-light steering at high road speeds.

Mounted on this out-of-the-ordinary chassis is the most delightful two-seater body imaginable. Beautifully comfortable, with adjustable seat and single dickey the finish is of the highest class and the style irreproachable.

The tout ensemble is one of the finest productions we have ever turned out from our famous Queen Street Showrooms. For a car of such distinction the price, £350, is extraordinarily modest.

May we send you further particulars ?

The Morris Garages
Head Office & Showrooms Queen Street
Oxford
Phone 942 Wire "Auto"

BELOW: Featured in the first issue (May 1924) of *Morris Owner* magazine, this ad is generally considered the basis for the claim that the octagonal badge was introduced in that month. *Author archive*

RIGHT: Cecil Cousins, a Morris Garages employee in the period before the Cecil Kimber era. Cousins was an important part of the MG story, working for the company until his retirement in 1967. *Sue Mott*

BELOW: A 1923 Raworth-bodied MG parked on Alfred Lane, Oxford. Six such cars were built. Sadly, none of the originals survive. Compare this image with the November 1923 ad in the *Isis*, shown on page 16. *Early MG Society*

commanded a certain amount of energy, organizational ability, and perseverance. The work involved in building complete motorcars was beyond the scope of the facilities in the Clarendon Yard, where the activities of the small team included Cecil Cousins, George Morris, and a new, initially unpaid member of the team, John Browne. The author interviewed Browne in the 1990s when he was approaching his own centenary.

As he recalled, their work at the Clarendon Yard, which often took place in the evenings, annoyed the managers of the workshop and the adjacent hotel. "We got kicked out of the Clarendon Yard, and the chassis work went to Longwall, where we carried on. Then a similar thing happened there—they decided that they wanted to close up earlier at night, closing at six—and the managers and so on wanted to lock up by seven. We were sure that Longwall had been tipped off by the Clarendon about us 'ruffians'!" Thus, in February 1923, the work moved to a garage in what was then known as Alfred Lane (the names Alfred Street and Alfred Lane were changed in 1926 to Pusey Street and Pusey Lane in honor of Edward Bouverie Pusey, a Victorian Oxford theologian).

Browne could remember the place clearly when the author interviewed him: "Kimber took over a little mews off Beaumont Street—a little dead end. It was a coach house to one of the big houses—I remember it had big, black-painted double doors opening outward. It had obviously been used in some car-related capacity before we arrived, since there was a rudimentary wooden 'test body' stuck up in the roof—which remained there whilst we were there, and was still in place sometime after we left." The chief occupation at Alfred Lane was to turn donor Morris chassis into Morris Garages Chummy models.

Meanwhile, Morris Garages' success with the Chummy body style seems to have led indirectly to the introduction of a factory-assembled competitor from the parent Morris business. This vehicle, the Morris Cowley Occasional Four, appeared in time for the 1924 sales season. With the obvious fact that these lower-priced cars were assembled for sale in one process rather than being reworked by a separate workshop, it was

Oliver Arkell, a scion of the eponymous family brewery based in Swindon, Wiltshire, shown in his brand-new MG sports car, which he acquired in August 1923. *Early MG Society*

clear that Kimber had some slightly unwelcome competition from rather too close to home.

This was a turning point. Kimber placed an order with the coachbuilder Raworth for six two-seater bodies to be fitted to Cowley chassis. Here was the evolutionary moment where, as it were, the Morris Garages sports cars first emerged on dry land: the initial differentiation that would lead to the first recognizable MGs. John Browne remembered the building process used for the six Raworth-bodied cars. The coachbuilder was situated at St. Aldates, just below and opposite Christchurch College. "There was a foreman called Pittam who gave us a lot of help in forming the body lines and structure."

At a retail price of £350, they would prove quite a tough sell, but nevertheless they found a few customers who were swayed by their novelty. Oliver Arkell, scion of his family's eponymous brewery in Swindon, lived just south of the Wiltshire town of Highworth. He was an early customer: His Raworth car, registered as FC 5855 on August 16, 1923, was painted in a bold color scheme of yellow and black. Many historians have come to regard this as the first MG, although Kimber rather muddied the water in terms of the debate. Although Arkell's car, which his father dubbed as "Oliver's Beetle," is long gone, a beautifully crafted replica, built by enthusiast Stephen Hiner, emerged in 2023 and is shown here. While these MG Super Sports Morris cars are generally regarded as the first MGs, the discrete car of this make did not materialize until 1927.

According to Browne, the converted cars were based on both Morris Oxford and Cowley chassis. "We'd have one in at the Alfred Lane lockup, take it down to Raworths and then finally go round to the Park End Street showroom and head offices. Cecil Kimber would then use it to drive round to the agents to drum up interest in his 'specials.' He might then sell that one to the distributor, who would order more, and so demand built up." We will return to the matter of distributorships later.

Soon after these initial sales, Jack Gardiner, a young salesman with Morris Garages in Queen Street, decided he would like a special car for his twenty-first birthday. He obtained a Morris Oxford 14/28 with special coachwork and, again, a striking color scheme. This car was completed and registered FC 6333 on March 13, 1924. The Gardiner car was another step on the road to what would become the MG style.

Another car was built for Billy Cooper, a well-known trial driver. This iteration sported

ABOVE: A car generally regarded by many historians as the first MG: a Raworth-bodied, Morris Garages–built Morris Cowley, acquired by its first owner, Oliver Arkell, in August 1923. The original car does not survive, but this stunning replica was built by MG enthusiast Stephen Hiner in 2023. Shown here in its public debut at Silverstone in June of that year. *Author archive*

RIGHT: The carburetor on the Raworth re-creation is of an early "sloper" type of SU unit. *Author archive*

a shiny aluminum-alloy body from Carbodies of Coventry, set off with blue interior trim and wings. The simple visual statement of this car must have been almost as striking as a chrome wrap might be on a modern car today. The taste might be questionable, but the visual impact was unmistakable.

By the time of the aforementioned May 1924 advert in the *Morris Owner*, Kimber had pulled a masterstroke. By lowering the rake of the steering column, subsequent cars at last eschewed the rather pedestrian, upright stance of the original donor Morrises. Early MG Society member and historian Christopher Keevill notes that when the steering wheel was lowered, to provide a lower line for the Bullnose MGs, the steering column was raised at the front by means of a special bracket. "This meant that a drop arm 50 percent longer than normal had to be installed. Consequently, the steering was much heavier on an MG than on a Morris. Salesmen were instructed to sell the MGs with high-pressure beaded-edge tires to help lighten the steering. No allowance was to be made if customers asked for high-pressure tires instead."

John Browne added that there were other special treatments:

> *In order to give the cars a more sporting, lower stance, the springs were modified too. They'd get two or three sets of springs from Cowley, take them to Grants at Princes Street, where a wonderful old blacksmith would reset them cold: he would make chalk marks on the floor, and by striking the springs with glancing blows from his hammer he would stretch the metal and reshape them, often doing this straightaway while I waited. He would then check them in a special wooden template box to ensure that the shackle bolts would fit.*

So the process of development and differentiation continued. But Kimber was clearly bitten by the sporting car bug, and his heart was set on what he would always refer to as his first sports car. Before we turn to that car, it's worth touching on some of the other experiments.

ABOVE: An elegant MG Super Sports open tourer from 1924. To some eyes, the bullnose front lent an air of "Junior Bentley" to these rakish cars. The Ace wheel discs were standard equipment, hiding the rather ungainly Morris artillery-type wheels. *Early MG Society*

LEFT: By way of comparison with the standard fare from Morris, this 1925 Morris Cowley four-seater tourer represents a fine, sturdy motorcar—but without an ounce of sportiness in its design. *Early MG Society*

ABOVE AND RIGHT: This 1925 MG 14/28 Salonette is owned by Mike Dacre. It's believed to be the only Bullnose of its type and period to have survived. Note the pronounced boat-tailed rear end. Kimber was a keen sailor and enjoyed adding nautical touches where he could. ***Early MG Society***

Partly in response to the Morris-built competition referred to earlier, Kimber tried a one-off Morris Oxford-based Chummy and a somewhat less sporting saloon with upright aluminum bodywork, built on the Morris 14/28 chassis, which was dubbed the "MG vee-front saloon." More promising would prove a four-seater Morris Garages car based again on the Morris 14/28, built broadly on the lines of the Vauxhall 30/98 (Vauxhall was still quite an upmarket marque at the time) with a "rakish body of polished aluminum," as noted by those who remembered it.

The story of that rakish car began when, early in 1924, Morris Garages in Queen Street was called on by Reg Brown, an employee of John Marston Ltd., makers of the Sunbeam motorcycles sold from the showroom. Brown

arrived in his highly distinctive one-off Morris Oxford 14/28, rebodied by Clarry Hughes of Birmingham and said to have been inspired by Brown's 30/98 Vauxhall. Christopher Keevill has established that the Brown car benefited from flattened springs, raked steering, wire wheels, and an open four-seater body, the latter made of polished aluminum with upholstery in red leather. Kimber was clearly entranced. By March 1924, he had readied the first MG Super Sports, also on a 14/28 chassis and with a bespoke body. Keevill reasonably speculates that this car may have been from the same Birmingham coachbuilder that Reg Brown had used.

In September 1924, Morris introduced a series of improvements to the Oxford, notably to the brakes and chassis frame. Kimber soon used these new base components to enhance his own Morris Garages creations, and the result was the definitive MG Super Sports model—better known to most enthusiasts and historians as the MG 14/28 Mark I. The two-tone body, doubtless inspired in part by the Billy Cooper one-off, offered a lower half polished in bare alloy, with the bonnet top and other upper surfaces painted, generally in red or blue.

To what extent these cars can be truly regarded as MG 14/28 or Morris Garages–modified Morris Oxford 14/28 is one of the many debates that vex early MG folklore. Certainly the contemporary vehicle licensing records at Oxford seem to record most pre-1927 cars as Morrises, although the MG octagon appeared on the door tread plates since at least the beginning of 1925. Keevill stresses that there was no octagon on the door threshold plate of the Raworth-bodied Super Sports of 1923, stating, "I have a threshold plate from one of those cars, given to me by someone who had owned it in the 1920s." The answer to this puzzle depends to some extent on who one asks.

A very brief exercise was a first foray into the world of the six-cylinder engine, taking as its basis a six-cylinder Morris Oxford introduced at Cowley with what was called the F-Type engine (not to be confused with the later MG F-Type). The engine suffered from vibrations that could cause the crankshaft to break. John Browne remembered that the unit "effectively comprised a Cowley block with two extra cylinders, and suffered from a sand trap in the front of the block, which caused the camshaft to overheat." This first six-cylinder MG was abandoned after fifty had been built.

BAINTON ROAD TO THE RESCUE

So successful had the MG Super Sports become in the meantime that sales were unshackled from the Oxford showrooms alone. January 1925 saw Morris distributor Stewart and Ardern offering the MG Morris Oxford Sports four-seater and before long had proclaimed their status as the sole London distributors of these MG models.

A 1925 MG Super Sports two-seater with rakish bodywork and Ace wheel disc covers—just the thing for the young man about town. *Early MG Society*

Old Number One again. Note the subtle MG badge on the left flank of the bodywork. *Author archive*

The growth in demand also showed that the existing production arrangements were no longer adequate, and Kimber was faced with finding a facility with more space.

The answer, for the short term at least, was to take over a couple of disused bays, about 160 feet (49 m) long and 20 feet (6 m) wide each, within the Osberton Radiators plant in Bainton Road, North Oxford. This factory was one of the businesses that William Morris had bought outright, keeping much of the management but reorganizing production in a way that saw output grow to meet his needs.

Production of the MG models began at Bainton Road in September 1925. The author was fortunate to meet and interview one of the new arrivals to the business at that time, Syd Purves, whose memories illustrate the nature of production, the workforce, and some of the key changes that set apart these MG models from the regular Morris models upon which they were based. The Bainton Road MG workforce was no more than a dozen strong. With Purves were the works manager George Propert; Cecil Cousins, in charge of engines; Jack Lowndes overseeing chassis work; George Hamilton, responsible for the paintwork; Jim Prickett, the electrician; George Morris, who had been a mechanic at Longwall; Albert Eustace, in charge of stores; Pat Wright, the test driver; Alec Hounslow; Frank King; and two other members of staff—a second test driver and a cleaner (probably Stan Saunders and Fred Hemmings, respectively). Hitherto largely an unpaid volunteer, John Browne soon joined the full-time paid staff at Bainton Road.

Browne told the author that the chassis delivery process comprised towing six chassis at a time. "They came from Cowley, over Magdalen Bridge, Long Wall, and Broad Street, all in the middle of town like a long snake: quite a thrill if you were the last man!" Occasionally, Browne claimed, the chain could be up to eight long, with just one set of trade plates (fitted front and back).

Working on chassis preparation under Jack Lowndes, Purves found himself helping to convert chassis at that same rate: six per week. Before he could start work, he had to buy his own set of three Whitworth spanners and a pair of small pliers. Every Monday morning, six chassis—usually 14/28 Oxfords, but occasionally Cowleys—would be delivered down the ramp into the first of the two bays at Bainton Road. Chassis modifications included changing the gearbox top cover and lever, which gave a more rakish alignment, while at the rear of the chassis the original Morris chassis springs were removed and sent off to be given a flatter profile. When refitted, this also lowered the whole body.

The engines, still mounted in the chassis, had their heads removed and polished, reassembled, and taken to the engine shed. There they were connected to the gas main, started with the aid of a little petrol in the carburetor, and run on the mains-fed Town Gas for about a week to break them in. The brakes were upgraded with a De Wandre servo, and André Hartford friction shock absorbers were fitted at the rear, which involved cutting away the Morris shock absorber brackets and bolting on new phosphor-bronze shock absorber brackets. Also, after late 1925, new wire wheels were installed, replacing the basic Morris units, which had actually been artillery wheels covered with Ace disc wheels.

Thus equipped, the now roadworthy chassis were sent to be tested with soapbox seats. If they passed muster, they were deemed ready for Carbodies at Coventry, where they would receive their bodywork. Christopher Keevill has established that the Hartfords were used on William Morris's instructions, "so as to use up all those in the stores that were left over after the six-cylinder F-Type project was abandoned."

KIMBER'S FIRST SPORTS CAR: OLD NUMBER ONE

Cecil Kimber entered a modified Morris Garages Chummy in the 1923 London to Land's End Trial, in which he qualified for a gold medal (Kimber elected to take the alternative gold cufflinks instead). Bitten by the racing bug and spurred on by the publicity, he resolved to come back, but next time with something rather more special. The one-off two-seater that resulted, dubbed in hindsight as "Old Number One," has long been the subject of confusion and debate. This is due in part to Kimber's insistence on referring to it as his first true sports car, and also because some early historians—no doubt befuddled by inaccurate Nuffield Organization publicity—erroneously stated its date of origin as 1923 rather than 1925.

In March 1924, Charlie Martin started work at Longwall and began dismantling and rebuilding a 1,548 cc Hotchkiss 11.9-horsepower engine. According to Keevill, "it was a Type CD overhead valve engine, introduced in 1920, of the type also supplied to Gilchrist Cars of Glasgow and to the Autocrat Light Car Co. of Birmingham." Later in the year Frank Stevens was tasked with taking a Cowley chassis and cutting it to Kimber's requirements. By New Year's Day 1925, plans had been laid to enter the new car in the forthcoming Land's End Trial. The chassis was fitted with a unique Carbodies body (built on March 13, 1925), and the finished ensemble was painted gray (its present red color stems from a subsequent postwar restoration). The car was registered on March 27, 1925, as a Morris Cowley Sports, Gray, in Kimber's name rather than that of Morris Garages.

The first week of April saw some rudimentary testing, one consequence of which was a fracture toward the rear of the chassis. This meant that Cecil Cousins became involved

Old Number One made an appearance at a major U.K. centennial celebration in summer 2023. A cutout of Cecil Kimber stands in the background. *Author archive*

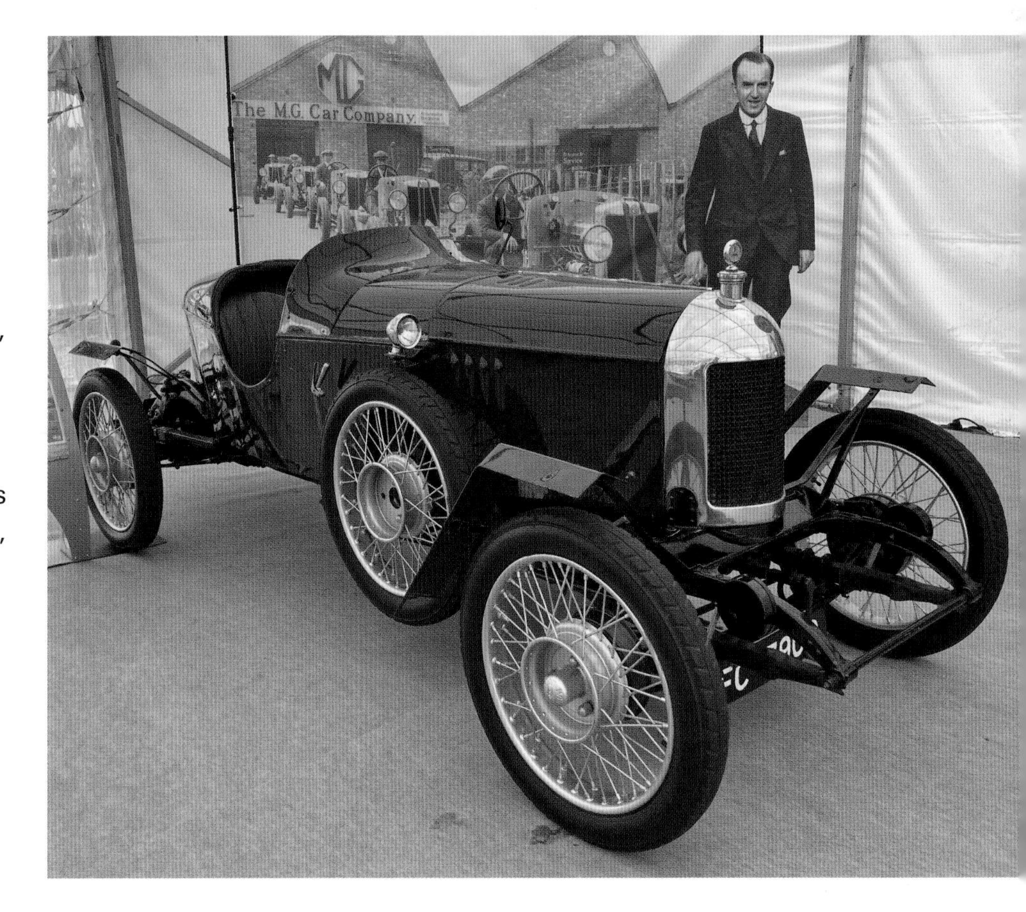

RIGHT: The familiar form of the SU carburetor. William Morris had acquired the business, which meant that most MG cars featured SU carbs over subsequent decades. *Carl Gotham*

BELOW: Kimber at the wheel of a flat-nose MG 14/28. *Early MG Society*

in the welded repairs just prior to the weekend of the race, on April 11–12. Kimber drove it with his friend Wilfred Mathews from the start in Slough to the final stages in the West Country, and was gratified when the car secured a gold medal in the Light Car Class. With this he had achieved the publicity in the motoring media that he craved. For the record, it may be noted that the entry, covered in the *Light Car & Cyclecar* of April 25, 1925, refers to Kimber's car as a "11.9hp Morris-Cowley." Neither Morris Garages nor the term *MG* received a mention. Some contemporary controversy surrounded the fact that, as the engine capacity was 1,548cc, it was slightly too large for the category (under 1.5 liters) in which it had been entered.

The success of the exercise notwithstanding, the car was unceremoniously sold off to Stockport Morris dealer Harry Turner, a friend of Kimber's, for £300. Less charitable people have suggested that Kimber may have wanted to get rid of the incriminating "oversize" engine! As MG's status grew in the following decade, though, a subsequent search in 1932 found it languishing in a Manchester scrapyard. Rescued, it was bought back by an MG employee for £15.

THE END OF THE BULLNOSE

The distinctive rounded shape of the bullnose radiator had been one of the hallmarks of both the Morris and the Morris Garages ranges. This changed in late 1926, when Morris made substantial changes to his mainstream models, including the appearance of a new flat-fronted

radiator along with a broader and more substantial chassis. It could have been a crisis for Kimber: The combination of the distinctive, rounded radiator and sporting coachwork had led some to consider the MG Specials as resembling pocket Bentleys. Inevitably, the Morris Garages Specials had to follow Cowley's lead, which meant the introduction of the so-called Flat-Rad 14/28, a design that later gave way to the similar MG 14/40. The alterations to the body to match the new chassis were designed by Hubert Noel Charles. The seasoned automotive engineer had joined Morris Motors in 1924, but began moonlighting for Kimber, often working in the evenings at the home Kimber shared with his artistically inclined wife Renee (whose given name was Irene), before finally being brought on as a full-time MG staff member four years later.

A "German silver" plate section was soldered over the base of the radiator grille of these early flat-front 14/28 MGs in an effort to make the grille look less tall and to offer a slight distinction from the Morris donor, but the round badges remained distinctly different from their octagonal counterparts; also, they still referred to Morris Garages. Full-depth doors made access much easier, while the two-tone color scheme was revised, with the highly polished aluminum lower panels replaced by an engine-turned finish. The standard Morris steering was replaced by the more effective Marles setup, surmounted by an elegant René Thomas sprung steering wheel. Magna wheels—wire wheels with large-diameter five-stud hubs— were fitted with modern balloon tires.

ABOVE: A 1926 season MG Super Sports, complete with "Calometer." *Early MG Society*

LEFT: A typically colorful contemporary advertising brochure for Morris Garages. The artist in this case was Leslie Grimes. Kimber also employed Harold Conolly for his graphics. *James Mann*

MG'S FIRST FACTORY: EDMUND ROAD

In May 1927, work began on a brand-new, purpose-built factory at Edmund Road, in the Oxford district of Temple Cowley. The design and construction represented a remarkable exercise that was fully funded by William Morris himself. Cecil Kimber gives the best account of building the factory in his response to a question following his presentation, "Making Modest Production Pay," given just seven years after the Edmund Road factory opened.

> *I was fortunate in having Lord Nuffield behind me, who did not draw profits out of the business. Practically the whole of our expansion has been out of profits. As an example of the extremely kind way in which Lord Nuffield does things, we first of all started producing in very small amounts in an unwanted bay of the radiator factory; then we required two bays, then six. I went to him and he said, "We shall have to do something about it. What do you suggest?" I replied, "I propose we spend £10,000 and build another factory." He said, "Where do you propose to build it?" I told him, and he said, "Right, go ahead." He did not see a plan or a specification, and we actually spent £16,000, and all out of the profits we made. This is the main secret of the present standing of our business.*

A cynic might note the repeated reference in those remarks to "profits"—and the cynicism would be justified, for Kimber's MG business record was not noted for a settled bottom line. Nevertheless, the story underscores the trust that Morris placed in his Morris Garages manager.

The Old Speckled Hen—a corruption of "owd speckled un," literally "the old speckled one." The bodywork of this MG 14/40 was by Gordon England, who was better known at the time for his Austin coachwork. *Author archive*

The new Edmund Road factory took six months to build, the contractors being the respected Oxford firm of Kingerlee & Company. According to John Browne, "Carl Kingerlee's son had been manager before [George] Propert for a short time, but he didn't prove suitable for the job and so left. His father's company was a building concern, and as old man Kingerlee and Billy Morris had grown up together, it wasn't too surprising that he got the job of building the new factory."

Cash-rich Morris added another new business to his growing personal portfolio in this period. The acquisition of Wolseley Motors represented a complete enterprise, including car manufacture as well as a highly regarded engine design.

Wolseley had grown to be one of Britain's largest car manufacturers after the war, thanks in part to the support of Morris's chief rival, Herbert Austin. When it teetered on the verge of bankruptcy in 1927, the two men, along with an anonymous third party—possibly General Motors—sought to buy the failing company. Morris doggedly outbid Austin in one of those cases where a wealthy, determined, stubborn man would not be beaten in an auction. With this, Morris, who also had substantially deeper pockets than Austin, gained control of Wolseley.

As we shall see, Kimber and MG soon reaped important benefits from this acquisition. A year before buying Wolseley, Morris had acquired S. U. Company Limited from its founders, the Skinner Brothers. The growing size of his personal portfolio thanks to these and other purchases brought him to the attention of the U.K. tax authorities, which threatened a "super tax" applied to the wealthiest British citizens. To head off this threat of a substantial tax bill, the company performed some creative consolidation in the summer of 1927: Morris Industries was formed in June and, a month later, Morris Garages Limited came into existence. Soon enough, the MG Car Company (proprietors: Morris Garages Limited) appeared, a further step toward the foundation of the MG company we know today.

On the evolutionary path; the badge on the top of the radiator grille is still essentially Morris in style, but the cut-out letters on the face of the grille tell us that this is now an MG. Note the "Calometer." *Author archive*

The service bay at the brand-new, bespoke factory built for MG at Edmund Road in North Oxford. The "Old Speckled Hen" can be seen second from the right. *Author archive*

It's clear that the MG Car Company was formed as a deliberate move to distinguish the car manufacturer and associated sales organization from Morris's general retail business.

By September 1927, just as the Edmund Road factory started production, the MG model range was firmly established as an entity, offering three basic models: the MG Sports Two-Seater, the MG Sports Tourer, and the MG Sporting Salonette. By this time Kimber had begun to stealthily grow his network of dealers beyond his arrangement with Stewart and Ardern.

The new cars were at last identifiable as MG models, replacing the earlier hybrids of the 14/28 Flat Radiator Super Sports. The new MG 14/40 Mark IVs were the first cars to be fitted with an MG chassis plate and car number. As explained by the late Robin Barraclough, one of the foremost historians of early MGs, "despite the Mark IV label, the 14/40 is not the fourth, but the first model of MG as a make in its own right."

Furthermore, Kimber's growing obsession with the octagonal badge was manifest in a significant number of these symbols cast into bespoke MG componentry. He used the MG badge wherever he could: It appeared on door tread plates and, through 1927, it was applied as cut-out letters in German silver to the flat mesh of the radiator grille. Later in the year, an enameled MG badge finally usurped the special Morris Oxford—MG Badge that had been used on earlier vehicles since the 14/40 Salonette. The basic grille of the 14/40 was carried forward from the Flat Rad 14/28, but that model's panel was here deleted, and instead a neat-shaped, body-color apron was fitted between the forward chassis legs.

John Browne, who was still involved at this stage, recalled the work on the line where the new MG badges were soldered onto the grille. First the standard Morris badges had to be removed:

> *The men would heat up the area around the Morris badge on the radiator top with a flame, moving round and round the badge very gently, then hook the badge off—often splashing a little solder on the radiator in the process. Any splashed solder would be carefully removed using the flattened end of a piece of copper pipe with the aid of the flame, followed by a very careful application of the finest grade of emery cloth. Then the new MG badge—initially the Morris Garages MG type and later the familiar MG badge—would be carefully placed, with the aid of the sucker used for grinding in valves, and soldered in place.*

According to Browne, it was not unknown for an MG badge to be applied to an otherwise standard Morris.

Meanwhile, Kimber continued dabbling with coachwork design, overseeing experiments with a fabric-covered timber body that saw light as the MG Featherweight Fabric Saloon. Its coachwork was by Gordon England of London, a business more commonly associated with the Austin Seven. Sales were poor, though, and the prototype—with its unusual mottled gold-flecked black finish—became known at the factory as "Old Speckled Hen." This was allegedly a reference to it being the "owd speckled 'un"—literally the "old speckled one" in Berkshire dialect. Whatever its name, it was used as a factory hack for several years.

A new face at Edmund Road at this time was unindentured apprentice Jack Daniels, a sixteen-year-old who joined right out of school. In later years, Daniels formed a key part of the team responsible for the Issigonis Minor and Mini. He told the author of the sights that stayed with him from his first day at work in Temple Cowley, witnessing the delivery of Morris chassis to be converted into MGs. "I was fascinated by how they did this," he remembered. "There was one tow vehicle at the front, drawing five wheeled chassis behind it, each with a man standing on a board on the chassis, steering with the wheel and operating the rod brakes; that was quite something to see." Working under the direction of Foreman John Bull, Daniels would set to work in order to remove surplus parts and prepare the chassis, with the engine being removed, unlike earlier efforts at Bainton Road.

The engines were then upgraded. As Daniels explained:

> *There were five chaps with benches; one would do the valves and polish the ports, another would polish the cylinder head combustion chambers, after it had been skimmed elsewhere to raise the compression ratio, another would carefully "lap" each individual piston into its cylinder bore, whilst the last would hand-fit the white-metal bearings. [Key changes to the chassis included upgrades to the Morris braking system] on the Morris, the brake mechanism*

A 1927 MG Mark IV Salonette two-door coupe, complete with occasional "dickey seat" at the rear, for passengers who didn't mind being exposed to the elements. *Author archive*

Triple-M Register
MG
OWNERS
MG
CLUB
MG
VINTAGE
MG
REGISTER
CAR
MG
CLUB
PM·3

used two concentric tubes—one tube worked the foot-brake, and the other the hand-brake. MG decided to split this, but cleverly used the same pieces—just two more brackets were required. This certainly improved the brakes noticeably—especially the adjustments, which were much easier.

Once the cars had been fully modified, with the engines reunited with their chassis, they were broken in on rolling roads using Town Gas, in much the same way as earlier models had been prepared. Daniels said that the team handled up to five cars a day. "The rollers would take a whole five-day week for the complete series of running-in operations—cars would be running all day Monday to Friday during the daytime, with Saturday morning left to square up the week." Then, at the end of the production process, each car would be taken out on a road test, followed as appropriate by rectification and if needed, a second road test. George Morris was in charge of rectification and testing, with Sam Nash as his senior driver. Both men were still working at MG forty years later.

Kimber would later recall that the first known race victory in an MG-badged car was recorded as a hundred-kilometer event on a brand-new concrete-surfaced racing circuit at the San Martín autodrome near Buenos Aires, Argentina. The race was won outright in an MG 14/40 at the modest speed of 62 miles per hour (100 kilometers per hour) by local Alberto Sanchiz Cires. It was an auspicious debut, even though in reality the car only really won by default, due to a misunderstanding of how many laps had been completed when the race finished. This obscure result would soon pale in comparison with MG successes only a few years later.

THE FIRST BESPOKE MG RADIATOR GRILLE

As we have seen, the first Morris Garages Specials were closely based on the Morris

OPPOSITE: With its substantial casing, shaped plinth for the new enamel MG badge, and vertical rib, the distinctive MG radiator grille became a hallmark for MG cars until the 1950s. *Author archive*

BELOW: A 1931 MG 18/80 Mark I Fabric Saloon, with distinctive black fabric body, including green wings and running boards. *Author archive*

An MG 18/80 Mark I Speed Model today. *Early MG Society*

models from which they were derived. In the beginning, they shared the same grille shapes as the Cowley originals, even if there were gradual changes in detail, such as the badges. When the donor cars changed their radiator grille shapes from the distinctive bullnose style to flatter radiators, Morris Garages naturally followed suit. Kimber felt this wasn't enough: He knew that, apart from bespoke coachwork, the style and shape of a car's radiator grille was one of the few distinctive features that made a car recognizable.

Sometime in 1927, work began on creating a new radiator grille for future models. The precise date is unclear, although changes such as Morris Garages' change of status to a limited company in July and the new factory at Edmund Road in September help to give an idea of the chronology. John Browne, one of the witnesses of the process, stated: "There was a general carpenter employed by Morris Garages at the business's new place at St. Aldates, opposite the police station by Speedwell Street—we all called him 'Chippy'—so I don't remember his name. Kimber told him to get a block of beech and mount it on a plinth. As soon as Chippy had squared the block up, Kimber drew lines on the wooden block in pencil, and then instructed the carpenter to chamfer the wood to the shape he had defined." According to Browne, Kimber made a cardboard template and used this to help refine the shape he was after.

A few days later, he invited some of the staff to inspect the progress. "He asked us what we thought, and so we fiddled with it using a rasp. To make it fit the new 18/80 chassis, we cut off the plinth and sat the

The M.G. Midget Sports

£175

The M.G. Car Company. Oxford.

The M.G. Car Company

December, 1928.

TO ALL MOTOR TRADERS.

Dear Sirs,

All M.G. Sports Cars are designed and produced by specialists in the manufacture of Sports Cars. They are assembled in the M.G. model factory at Oxford which is devoted exclusively to this purpose.

The engines—which are built in Europe's finest engine factory—and most of the components, are special and exclusive to the M.G. Sports Car. Each car is treated individually from the start, and an extraordinarily high standard of efficiency is thus obtained. This results in an almost total lack of after-sales service required and thereby the conservation of your profit.

Judging by certain articles concerning M.G. Sports Cars which have recently appeared in the press, it seems that those responsible still do not appreciate the fact that M.G. Sports Cars are an entirely distinct production. In order to remove any doubt from the minds of British Motor Traders we repeat that M.G. Sports Cars are not ''hotted up'' or sports editions of any standard car, but are designed and built in every detail as Sports Cars and represent the entire output of the factory.

We should appreciate the opportunity of showing you round the M.G. works.

Yours very truly,

THE M.G. CAR COMPANY.

Cecil Kimber

General Manager.

LEFT: A very early brochure, highlighting the low cost of the new MG Midget, with accompanying letter from Cecil Kimber; both date to December 1928. Kimber encourages motor traders to come and visit the works, then still at Edmund Road, Oxford. *James Mann*

BELOW: Factory photo of an MG M-Type Midget, showing its diminutive proportions. With the top up, it was notoriously difficult to see clearly through the tiny windscreen. *Author archive*

block onto the chassis. However, we found that the wooden 'grille' was far too high for the bonnet, and so we had to cut some off. Unfortunately, we cut too much off, and so a bit had to go back on again."

Browne remembered that several people took part in the process, eventually working out the shape Kimber wanted.

"The block was taken down to Cudds," he continued, "where a tinsmith made a metal shell—and very nice it looked too. I went down on a chassis to collect the shell, wrapped it up in a sack and brought it back to Edmund Road, where we proceeded to fit it onto the prototype 18/80 chassis."

Osberton Radiators of Bainton Road (dubbed Morris Motors Ltd., Radiators Branch after June 1926) became involved, and the radiator core was largely the responsibility of Ron Goddard. The first public appearance of Kimber's new grille came with the announcement of the 18/80 in August 1928. It was the same basic shape used for the next three decades.

SMOOTHER ENGINES—THE MG SIX

As noted earlier, Cecil Kimber had worked at E. G. Wrigley before starting at Morris Garages. He knew Wrigley's general manager, Frank Woollard, a pioneer in the techniques of mass and flow production who was eventually brought on at Morris Engines in a senior role. Ever the clever networker with an eye for moving MG into new market sectors, Kimber managed to get Woollard and his designer, Arthur Pendrell, to agree to create a modern overhead

An MG 18/80 Mark II with coach-built four-door saloon bodywork. A sophisticated product from the Morris stable. *Early MG Society*

camshaft 2,468 cc engine—a design to be shared with the new Morris Isis.

The engine was singled out for description by Miles Thomas in *Out on a Wing*: "Morris never gave this engine his blessing, and his only half-hearted enthusiasm caused it to be mounted in three particularly unsuitable forms of chassis. First it was put into an elongated Morris Oxford chassis and called the Morris Major. The lengthening of the chassis frame made it whippy, particularly at the front end, and although the car, being light, had a good speed performance, it was an absolute bitch to hold on the road."

All the Morris cars that used this "light six" proved to be commercial failures, but Kimber was undaunted. He acquired a Morris Light Six fabric saloon and asked Cecil Cousins to draw up a bespoke chassis frame. This would become the first-ever unique MG chassis. The drawing board used was a sheet of plywood—Morris Garages had yet to invest in a proper drawing office, though that would soon change.

Work continued on improving the Light Six engine, which received a new block to accommodate twin carburetors mounted beneath the exhaust manifold. A package of changes delivered a smooth, quiet-running power unit. The prototype of the MG Six—more correctly the MG 18/80 Mark I—began to take shape in December 1927, the objective being to have the car ready for the 1928 Motor Show at Olympia. The track was kept at the same four-foot width as the Morris Six, owing to retention of the Morris axles, but the chassis itself was much different, even including some stylish MG scuttle bracing brackets that would be concealed under the bodywork on the finished car.

In early 1928, the MG Car Company (proprietors: Morris Garages Limited) was announced publicly. Here at last was the manifestation of MG as a carmaker in its own right. The new MG 18/80 was featured in the *Autocar* on August 17, 1928, making its debut at the Olympia Motor Show alongside another newcomer, the MG Midget. For Kimber, the bigger car was where he saw MG's future, as a kind of junior Bentley with some of the same refinement but at a lower cost. As it turned, he may have been surprised, along with everyone present, at the success of the other car shown at Olympia.

ORIGINS OF A LEGEND: THE MG MIDGET

The Morris Cowley and Oxford ranges were generally sold to the moderately affluent middle class, which had grown in number, status, and income throughout the 1920s. The

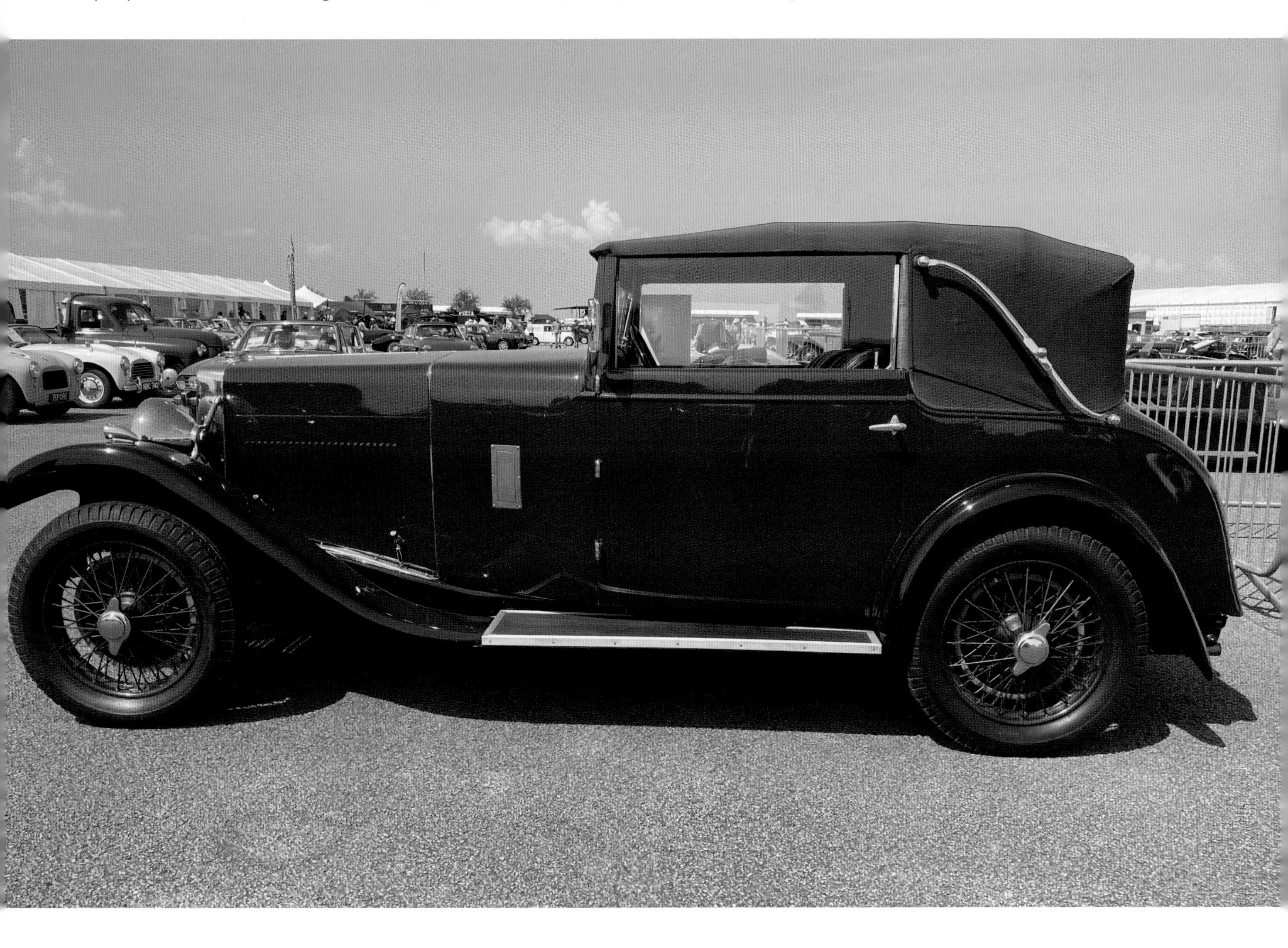

An elegant line was achieved with this Carlton-bodied MG 18/80 Mark II, which has been owned for many years by Ron Gammons. *Author archive*

MG
LS24
4

BLOCKLEY

PREVIOUS PAGES: The earlier MG M-Type Midgets had traditional fabric bodies stretched over ash framework. Note the semi-matte appearance of the rear tub of this early car. The vertical plate fixed on the front is a device used for time trials—a sign that this little sports car is still well used. *Cathelijne Spoelstra*

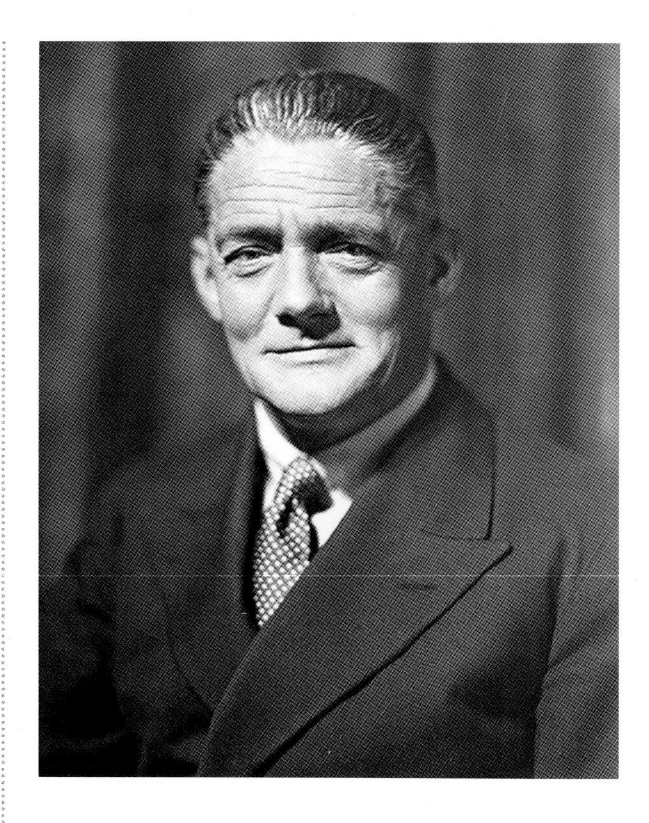

RIGHT: William R. Morris was first knighted in 1929, became ennobled (as Lord Nuffield) in 1934, and then became a viscount in 1938. He is probably better known by the general public for his philanthropy in the world of medicine. *Author archive*

fancy products offered by Morris Garages could easily end up costing twice as much as the donor car, so they inevitably appealed to those with greater disposable income. Such buyers were either unable or unwilling to make the leap to the more expensive tier, the upper crust of owner-driver motoring exemplified by the bigger Sunbeams and Bentleys. It was the middle-class, in-between market that Kimber had always coveted. Now, if Wolseley could provide more staid luxury, as far as he was concerned, MG could provide the sporting pizzazz to justify a higher retail price.

Another, parallel trend in the automobile industry was taking place, however: A new class of more affordable motorcar had emerged, typified by the popular Austin Seven, launched in 1923. This car, a more basic entry in the world of motoring, was certainly a step up from the privations of a motorcycle-sidebar combination. Not to be outdone, Morris entered the fray with his 8-horsepower Minor, a solid seller helped by Morris's aggressive pricing policy at a time when sales of Morris Oxfords and Morris Cowleys were declining due to increased taxation.

A surprising feature of the baby Morris was its Wolseley-designed overhead-camshaft four-cylinder engine, which offered latent tuning potential. Long before coming under Morris ownership, Wolseley had gained useful experience by building Hispano-Suiza engines under license during World War I, and the relative sophistication of the Morris Minor engine design owed much to this, even if in due course the model would move over to more mundane powertrains.

The Minor was unveiled in August 1928 in time for the autumn show at Olympia. Less than a month later, the *Motor* floated a rumor that the Minor would soon be joined by a new offering, the "Morris Midget, to be built by the MG Car Company of Oxford." Kimber had in the meantime taken a prototype Minor to Carbodies and had them fit it with a new open two-seater fabric-covered ash-framed sports body, featuring a distinctive boat-tailed rear end. Back in Oxford, the Minor's suspension was lowered, its steering column raked flatter, Hartford shock absorbers fitted, the brakes upgraded to 8-inch (20-cm) drums, and MG hubcaps fitted to the Morris Minor wheels. Yet much of the open car remained fundamentally a Morris Minor, as an echo of the first Morris Garages Specials. The Wolseley-designed 847 cc four-cylinder engine gave the Minor a modest output, but for MG's needs, a sportier tune was just what they wanted.

At Olympia in 1928, a great amount of what we would today call buzz was generated by the MG 8/33 M-Type Midget; this was the first appearance of the Midget name, which would go on to become central to the MG story for the next fifty years. A retail price of £185 made the diminutive sports car attractive, along with the fact that it weighed barely ten hundredweight (just over 1,100 pounds or 500 kilograms), yet produced 20 horsepower. By the following spring, production got underway at Edmund Road, while the bodies were initially mated with the running chassis at Leopold Street. The *Autocar* published an early test in which it memorably stated that "sixty or sixty-five miles an hour are not adventure but delight, acceleration is very brisk, altogether an extraordinary, fascinating little vehicle."

MG ENHANCES THE SIX: MG 18/80 MARK II

For Kimber, the MG Six was just the first step in his ambition to move subtly upmarket. A new, more sophisticated (and more expensive,

William Morris and His U.K. Honors

The first public acknowledgment of Morris's successes was his receipt of an OBE (Order of the British Empire) in 1918 for his work during the war. He became Sir William Morris, Baronet, in 1929, then Baron Nuffield, of Nuffield in the County of Oxford, in 1934, which entitled him to be known as Lord Nuffield. The announcement of the latter title appeared in the *London Gazette*:

Whitehall, January 15, 1934

The KING has been pleased, by Letters Patent under the Great Seal of the Realm, bearing date the 13th instant, to confer the dignity of a Baron of the United Kingdom upon Sir William Richard Morris, Baronet, O.B.E., and the heirs male of his body lawfully begotten, by the name, style and title of BARON NUFFIELD of Nuffield in the County of Oxford.

He was later elevated to the title of Viscount Nuffield, again of Nuffield in the County of Oxford, in 1938:

Whitehall, January 28, 1938

The KING has been pleased, by Letters Patent under the Great Seal of the Realm, bearing date the 24th instant, to confer the dignity of a Viscounty of the United Kingdom upon William Richard, Baron Nuffield, O.B.E., and the heirs male of his body lawfully begotten, by the name, style and title of VISCOUNT NUFFIELD, of Nuffield in the County of Oxford.

Following this honor, he was named Knight Grand Cross of the Order of the British Empire in 1941 and appointed to the Order of the Companions of Honor in 1958. He was also accorded many honorary titles arising from his many philanthropic endeavors.

by £100) model was soon in development—the Mark II—with a stronger chassis that was four inches (ten centimeters) wider, a four-speed gearbox, and better brakes. And its sufficiently distinct chassis was named "Type A." The factory demonstrator was registered as WL 9232, and four days later it was competing at the 1930 Land's End Trial, with H. E. Symons of the *Motor* behind the wheel. The car's first owner subsequently took it on a world tour in 1931.

The new Mark II did not directly replace the earlier model, but the two were sold together over a two-year period up to 1931, when the Mark I was finally discontinued. During 1929, the Mark I was improved, with uprated cable-operated brakes, and in 1930 a speed model was introduced. A guaranteed 80 miles per hour (129 kilometers per hour) was claimed for the speed model, and a Mark II equivalent was also offered, but this again suffered with the added weight of the chassis, which detracted from the good acceleration that had been an admirable trait in the Mark I. As its power output was not increased, the Mark II generally proved harder to sell. Over their lifetimes, the Mark I sold 501 and Mark II 236 units.

By the summer of 1929, the new Edmund Road factory was already bursting at the seams with 14/40, 18/80 Mark I, and Midget production. A new decade was dawning, and with it an exciting new chapter in the MG story.

2

Midgets, Magnas, and Magnettes

DEVELOPING THE MIDGET

The early success of the M-Type Midget almost caught Cecil Kimber unawares: While his prime interest remained the task of moving MG upmarket by creating cars for the sporting gentleman, the Midget existed at a lower end of the spectrum. Kimber need not have worried. Being a pragmatist, he soon realized the attractions of the smaller sports car line, which especially benefited from its tiny jewel of an engine.

His design team was key in making the Midget run. Chief designer H. N. Charles, who had started from humble but promising beginnings, achieved such remarkable power outputs from the same basic structure that the little MGs would eventually capture an international audience, including Germany's resurgent engineering powerhouse.

Although the Edmund Road works were still only two years old in 1929, it had already become clear that demand would outstrip capacity. A search of the area revealed the old Pavlova leather works at Abingdon-on-Thames, a few miles south of Oxford, a substantial portion of which was vacant and available for rent. Established by Robert Fraser in 1912 and expanded during World War I to supply leather flying helmets and other equipment, the factory had fallen all but silent as demand for such equipment contracted with the war's end. Although the leather works were still operating, many of the wartime buildings lay unused. In fact, MG had been storing cars there for some time. Syd Enever, who would go on to great things in due

OPPOSITE: Inside the Abingdon Pavlova MG works. This 1930 photo shows the crossover between the older 18/80 sixes and the newer 8/33 MG Midgets. *Author archive*

ABOVE: Cecil Kimber and Lord Nuffield arriving for the celebratory luncheon to mark the opening of the new MG factory at the Pavlova Works. *Author archive*

RIGHT: A 1931 advertisement for the MG Midget. *Author archive*

course once MG moved to its new location, recalled the trips back and forth as the leather works were used more regularly.

George Gibson and Jack Daniels were the first to move to the factory in Abingdon in the autumn of 1929. "The clearing of the site was still underway, and we were put in an office downstairs in what was later Larkhill House," Daniels recalled. "And for a time, the builders had an office there right next to us."

Gradually, other staff began to move across to Abingdon. "Propert and Cousins and some of the secretaries moved into the downstairs portion of a newly built office block inside the factory proper, and we moved into a drawing office upstairs," said Daniels. On January 20, 1930, Kimber hosted an inaugural luncheon at the new works. The event featured Sir William Morris as the principal guest. In the next issue of the *Autocar*, it was reported that "Sir William Morris, Bart., made a characteristic and vigorous speech on the outlook for the motor industry and MG sports cars in particular. . . . Sir William paid warm tribute to the keenness and loyalty of the staff and workmen engaged in the production of the cars."

KIMBER'S JUNIOR BENTLEY

The ultimate evolution of the six-cylinder marks was the Mark III—the B-Type 18/80 Tiger, Tigresse, or 18/100. This last name came from abortive attempts to coax a reliable 100 horsepower out of the upgraded powertrain: the roadgoing version of the 18/100 barely achieved 83 brake horsepower; with further development, this was pushed up to 96. The engine had a crossflow head, the first of its kind in an MG. There was a new crankshaft, camshaft and pistons, and the added sophistication of dry sump lubrication.

Details of the new Tigresse first appeared in *The Motor* in February 1930. It looked like a kind of junior Bentley, which was what Kimber was clearly after. The exhaust had a sporting fishtail, and there were cycle wings, louvred chassis panels and a leather bonnet strap. Weight was again a handicap, though, as was the optimistic price of £895. Jack Daniels was responsible for drawing up the substantial stay between the headlamps, designed as an aerofoil section.

The MG Sports
"The standard by which other sports cars are judged."
—The Morning Post.
The 8/33 M.G. Midget Sports Two-Seater, £185
Safety Fast!
"The FIRST M.G. sports car was built in 1925. In its early days it was a real sports car; to-day it sets the standard by which other sports cars are judged. It is a fine tribute to the designers, that within such a short space of time this car has placed itself in the forefront so prominently that this year they have outnumbered any other make in amateur sporting events. Naturally the enthusiastic motorist selects the mount which he thinks is going to serve him best to gain a cup or a gold medal, and it has been proved beyond doubt that those who have entered an M.G. have chosen wisely."
—The Morning Post.
ISSUED BY THE PUBLICITY DEPARTMENT OF THE M.G. CAR COMPANY, ABINGDON-ON-THAMES

TOP: There is little doubt that the MG Six held its own as an elegant motorcar in its day. This example is a 1931 Weymann-bodied MG 18/80 Mark II two-door four-seater coupe. *Author archive*

ABOVE: The MG 18/100 was an imposing beast. Cecil Kimber considered it a design that could move Morris-based products beyond the commonplace. *Author archive*

The debut of the big-racing MG 18/100 in the 1930 Brooklands Double Twelve was a failure, but the same could not be said of the Midgets at the same event. The Hon. Victoria Worsley began racing in 1928, having won a car through a horse-racing bet. She finished the 1930 Double Twelve in twentieth place overall and seventh in class. *Author archive*

The 18/100's first appearance was less than spectacular, however. The engine seized after running its bearings barely two hours into its Brooklands debut, while the buzzing Midgets were running rings around their opposition. The success of the Midgets at Brooklands, the highest having finished third in class and fourteenth overall, and the attainment of a team prize, would prompt Kimber to announce a limited series of replicas of the race cars, claiming that they were identical in every respect to those that raced at Brooklands. *Motor Sport* drove Victoria Worsley's actual race car soon after the race, proclaiming that its high-speed road-holding gave the Midget what they called a "big car feel."

Meanwhile, Kimber blamed the Morris Engines branch for the failure of the 18/100. Fifteen years later, he wrote about the incident in a paper that he never got to present: He explained that, although he had managed to get Morris Engines interested in the idea of a special racing engine, they refused to listen to MG's suggestion that the crankshaft be fully balanced. "We did timidly suggest to the Engine Works that this might be necessary but we were told to run away and look after the chassis we were building; but the fact remained that owing to the crankshaft being unbalanced, the throw-out loads at high speed were so great that when eventually taken down after the race, the crankshaft main journals were actually blued from the heat that had been generated."

In the event, just five Mark III models were built. Kimber had hoped for a run of at least five times that. It was clear that the bigger Marks had nearly run their course. The Midget and other models derived from it would cement success for MG in the coming years.

MIDGETS AT ABINGDON

Despite Kimber's initial aspirations for the 18/100, production at the new Abingdon works was unsurprisingly soon dominated by demand for the exciting new M-Type Midgets, with up to thirty a week soon being built. The Mark I and II Sixes were still in production alongside them, but their star was on the wane; stocks of these cars were still around until 1934, long after their manufacture had ceased.

Another step change occurred in the late summer of 1930: The MG Car Company Limited was formally established on July 31. The new company issued nineteen thousand shares, all but five of which were bought by Morris Industries Limited, the balance retained by Sir William Morris, Morris's solicitor, his secretary, his accountant, and one by Kimber.

The previous month had seen the arrival at Abingdon of John William Yates Thornley, a young enthusiast who had chosen an MG Midget for his twenty-first birthday. Thornley would go on to devote his career to the development and preservation of the Abingdon works.

The summer of 1930 also saw the result of a concerted effort to extract more power from the Midget's minuscule engine, which had become something of a recurrent theme. We saw earlier how racing success at Brooklands over the weekend-long "Double Twelve" event had subsequently led to a short run of race-car replicas, dubbed the "12/12M." Evolution enhanced the M-Type: A more substantial pressed-steel body replace the earlier fabric-covered panels, and worthwhile improvements were made to the brakes and other running gear. Kimber was keen to support the closed saloon–type body, and a few MG Midgets were offered as two-seater Sportsman's Coupés, one of which found its way across the Atlantic into Edsel Ford's collection.

The M-Type was clearly a remarkable entrant into the small sports car sector, appearing for its final motor show at Olympia in 1931. Its Morris Minor chassis was a limiting factor, however: Having developed bespoke MG chassis for their larger models, Kimber and Charles were hardly content with compromise for the smallest.

Inspiration came from a neat French sports car, the Rally, with its light engines and a neat, straight-through chassis frame. Kimber got hold of one and immediately oversaw work to design a new Midget chassis under the project code of EX115, described as "Frame and Suspension—Midget 1931." Before long, however, the mission shifted to the creation of the first MG record breaker.

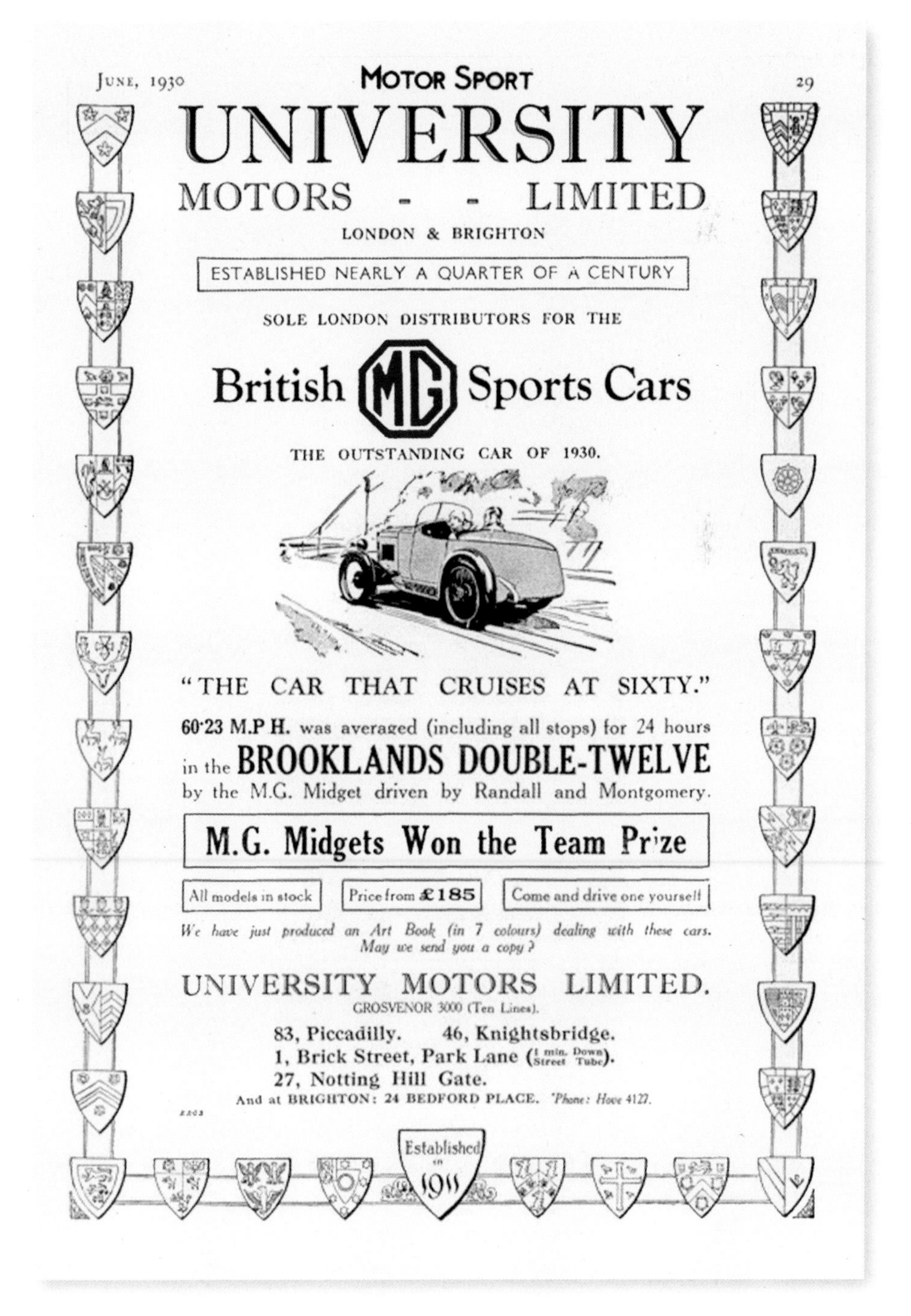

University Motors Limited was the main London agent for MG at the time. Their ad in the June 1930 issue of *Motor Sport* magazine celebrates the outcome of the recent Brooklands Double Twelve. *Author archive*

High Speed
MG
Service Van
The M.G. Car Company
Pavlova Work
Abingdon on

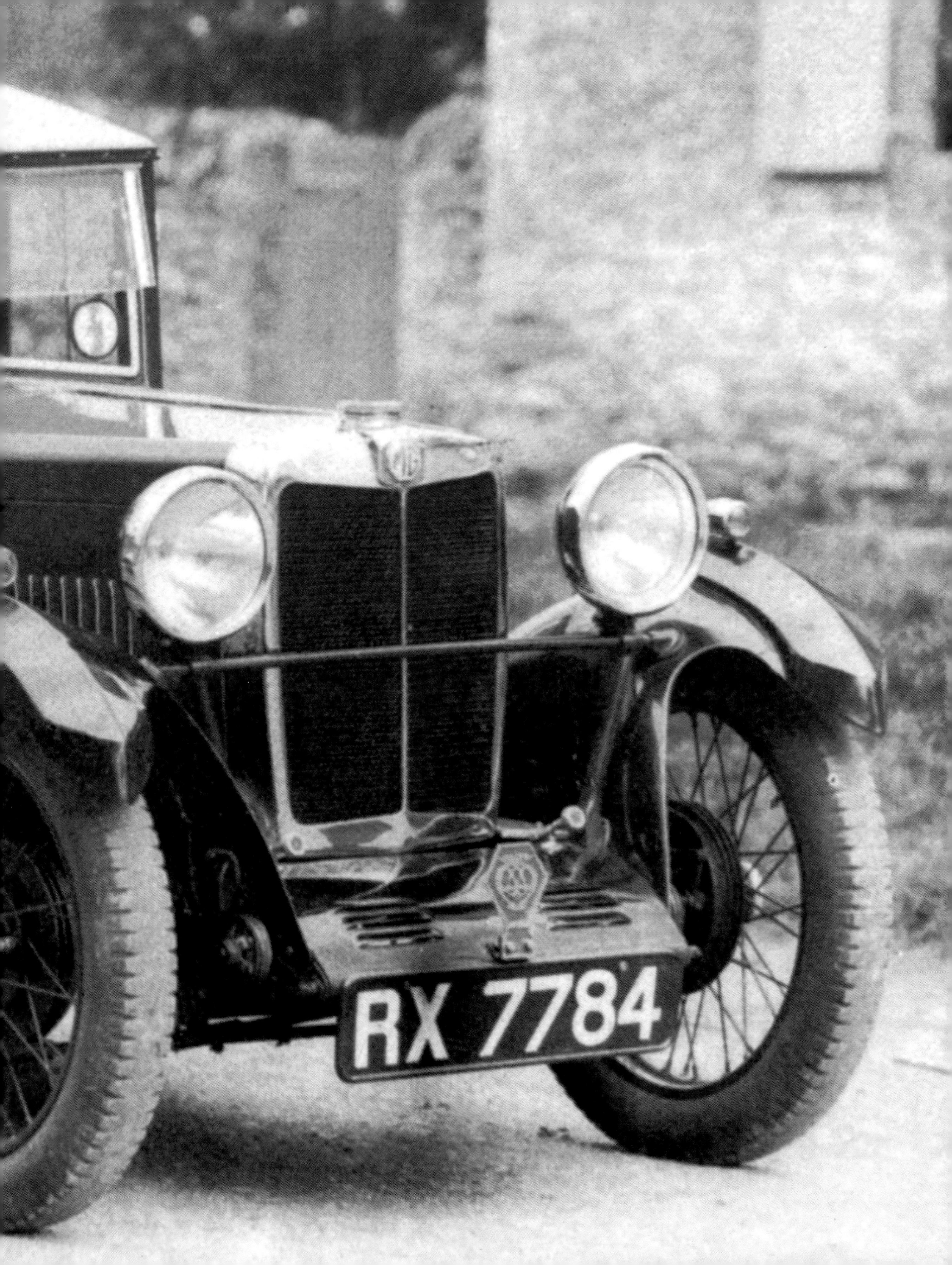
RX 7784

SAFETY FAST AND IMPROVEMENT OF THE BREED

PREVIOUS PAGES: This one-off MG High Speed Service Van, from 1931, was more of a gimmick, but it gained a fair amount of publicity. *Author archive*

Every business worth its salt deserves a good motto, a memorable phrase or slogan that people will associate with the product. Cecil Kimber tried "It Passes—and Surpasses," which was hardly inspiring. In due course he switched to "Faster Than Most!" He was less than thrilled when he saw that an advertisement had been doctored to read "Faster than most *bicycles*." At the same time, an article in the *Motor* of March 12, 1929, helpfully suggested that MG could consider "Safety with Speed" as a slogan. They were not far off, as it turned out.

Ted Colegrove, MG's first publicity manager, explained to his successor, George Tuck, how MG landed on its most famous slogan. Tuck in turn related the tale to the great MG enthusiast Norman Ewing. "Ted Colgrove was driving through Oxford—probably in October or November 1929—behind a new bus. A warning triangle was painted on the back next to a slogan, 'Safety First!,' which alerted drivers that it could stop more quickly than old buses thanks to brakes fitted to all its wheels. This gave Ted an idea: if it was changed to 'Safety Fast!' it would make a great slogan, and he rushed back to tell Kimber." It seems that the idea came just at the right moment, as Kimber was fuming at the bicycle comment, and "Safety Fast!" arrived as MG's famous marketing slogan (from 1959 it was even used as a magazine title). We know from contemporary magazines that "Faster Than Most!" was still in use in September 1930 (appearing in the September 5, 1930, issue of the *Autocar*), but "Safety Fast!" appeared in the *Morris Owner* in the same month.

MG
Safety Fast!
The M.G. Sports Cars
The New 18/80 M.G. Six Sports Mark I. Speed Model £525.
New Season's Announcement.
8/33 M.G. Midget Sports
Two-Seater (Triplex Glass), £185
Sportsman's Coupé (Panelled), £245
8/45 M.G. Midget Sports
"Double Twelve" Model - - - £245
18/80 M.G. Six Sports Mark I.
(With Chromium Plating, Triplex Glass and Servo-operated Brakes)
Two-Seater - - - £520
Tourer - - - £525
Sportsman's Salonette - £565
Four-door Saloon - £580
Speed Model - - £525
18/80 M.G. Six Sports Mark II.
Two-Seater - - - £625
Tourer - - - £630
Sportsman's Salonette - £660
Four-door Saloon - £670
Salon de Luxe - - £699
WRITE FOR OUR NEW ILLUSTRATED CATALOGUE.
ISSUED BY THE PUBLICITY DEPT. OF THE M.G. CAR COMPANY, PAVLOVA WORKS, ABINGDON-ON-THAMES

BELOW: One of the earliest instances of the famous MG "Safety Fast!" slogan: a September 1930 advertisement in the *Motor*. *Author archive*

Before long, the products from the new Abingdon works would prove worthy of the new slogan, as MG moved onward from the success of the M-Type. First there was a new record breaker that featured a highly developed version of the engine; its capacity was reduced to 743 cc to allow it to compete in the under-750 cc H class, and it was fitted with a Powerplus supercharger.

This unique car, EX120, came about after Kimber met with British record holder Captain George Eyston and a fellow Cambridge graduate, James "Jimmy" Palmes. They wanted to gauge his interest in achieving 100 miles per hour (161 kilometers per hour) in a car with a tiny engine.

EX120 was shipped out to the Autodrome de Montlhéry near Paris in December 1930 with the intention of capturing records, but initially the magic "ton" eluded them. The car was returned to Abingdon, where the news came that Eyston's rival, Sir Malcolm Campbell, was planning an attempt at 100 miles per hour (161 kilometers per hour) in an Austin Seven while he was in Florida. As far as MG was concerned, the gauntlet had been thrown! On February 6, 1931, Campbell managed 94 miles per hour (151 kilometers per hour) in the Austin, so clearly time was of the essence. The MG returned to Montlhéry and, after expending much midnight oil and hard work onsite, the EX120 achieved its mission as the light faded on February 16.

Kimber sustained the excitement stirred up by EX120 through the creation and marketing

LEFT: Captain Eyston at the wheel of EX120 at Montlhéry, France. *Author archive*

BELOW: Kimber capitalized on the successes of EX120 with this new model, the MG C-Type, which used the new chassis and a low, sleek body with an aerodynamic nose cowling. *Author archive*

of a new turnkey race car, the C-Type, more formally known as the MG Montlhéry Midget Mark II. The engine—whether supercharged or not—was a 746 cc version of the M-Type. The bodywork was made to resemble that of EX120 with a double-humped scuttle (the first time this feature appeared on an MG) and a sleek nose nacelle. Jack Daniels in the MG Drawing Office helped develop the engine under project code EX125. Still proudly thinking of the M-Type's Double Twelve success, Kimber resolved that the C-Type would be ready for the 1931 race over the weekend of May 8 and 9. By now a seasoned publicist for MG, Kimber arranged for detailed articles to appear in the enthusiast press at the beginning of March, providing ample time for customers to come forward. In the event, fourteen cars were ready in time for the race. Of those, seven finished, five of them taking the top places in order. The winner was Frederick Charles Gordon Lennox, the Earl of March, with Syd Enever serving as his riding mechanic.

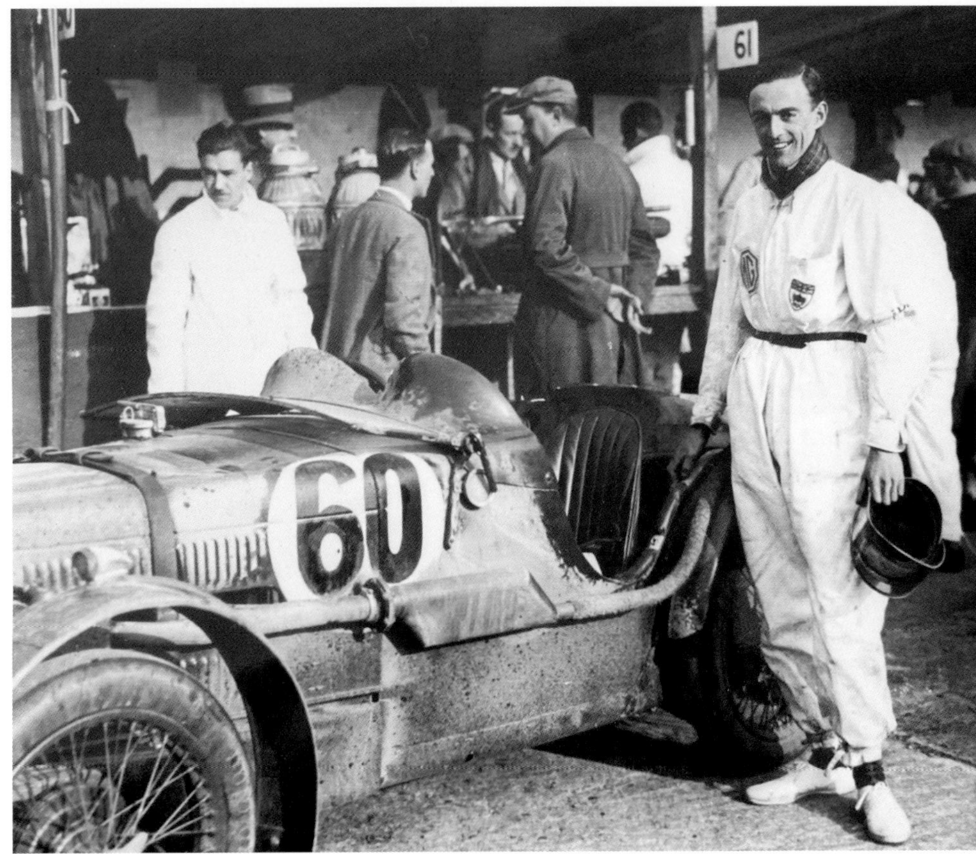

Although EX120's primary mission was over, it continued to take other records throughout 1931 until it expired in flames, at Montlhéry again, on September 25. Before the fire could engulf the car, Eyston managed to jump out while it was still rolling. A passing Citroën test driver rescued him and took the burned and shaken Briton to hospital.

THE MAGIC MIDGET

By this time, a second record breaker had appeared: EX127. This more specialized offering was dubbed the "Magic Midget." Reg Jackson took it upon himself to make a scale model at home, with a shape inspired by various larger record breakers, then brought it into the MG works, where Cecil Kimber gave it his blessing. Jack Daniels was tasked with much of the drafting work on the new record car, which boasted an inclined driveline that allowed the driver to sit lower in the cockpit despite it being a single-seater. While EX120 had secured the "ton"—100 miles per hour (161 kilometers per hour) —the objective with the Magic Midget was the next target, two miles a minute, or 120

OPPOSITE: EX120's demise came when it burst into flames while Eyston was driving it around the circuit at Montlhéry. Thereafter, Eyston took to wearing an asbestos racing suit (before the inherent dangers of this material were fully appreciated). Here Eyston and his business associate, Jimmy Palmes, survey the remains back at Abingdon, following Eyston's recovery. *Author archive*

ABOVE: Frederick Charles Gordon Lennox, the Earl of March ("Freddie" to his friends), with Syd Enever, acting as his mechanic, alongside the Earl's C-Type racing Midget in the 1931 Brooklands Double Twelve. *Author archive*

42
42

ABOVE: Norman Black (accompanied by MG mechanic Frankie Tayler) in his supercharged C-Type. With this car he won his class in the 1931 Royal Automobile Club International Tourist Trophy Race on August 22, 1931, at the Ards Circuit, Belfast. *Enever Family Archive*

RIGHT: A simple MG ad that got straight to the point. *Aviation Archive*

MG
MG
MIDGET
AGAIN
Double-Twelve
Irish Grand Prix
And NOW the Ulster T.T.
A Power Plus Supercharger was fitted to the winning M.G. Midget.
(Subject to official confirmation.)
ISSUED BY THE PUBLICITY DEPT. OF THE M.G. CAR CO., LTD., ABINGDON-ON-THAMES.

miles per hour (193 kilometers per hour). As Eyston was still recuperating, his associate Ernest Eldridge took the new car to Montlhéry in October 1931 and coaxed it to 110.28 miles per hour (177.5 kilometers per hour) over three miles (five kilometers) before a radiator burst stopped further efforts.

In December, Eyston felt able to drive the car—now wearing an asbestos suit as personal insurance—but the best he could achieve three days before Christmas was 114.77 miles per hour (184.7 kilometers per hour). A sudden change of international motorsports regulations meant that new records would henceforth be based on a pair of runs in opposite directions. The Montlhéry management were not prepared to close their track to facilitate such two-way runs by Eyston, and Brooklands was closed for winter maintenance. On short notice, then, the crew went to Pendine in South Wales in January 1932 to run EX127 on the seven miles (eleven kilometers) of sandy beach there. There were technical problems with the long cables lying in the salt water and, although one run was timed by stopwatch at 122 miles per hour (196 kilometers per hour), it was found that the RAC equipment's tracer pen had run out of ink!

By the time the equipment was fixed, the tide was turning and Eyston had to make do with a best recorded run at 118.39 miles per hour (190.5 kilometers per hour). Impressive, but not the hoped-for two miles (3.2 kilometers) a minute. Fitted with the new AB cylinder head developed for the Mark III C-Type, the car under Eyston's control eventually achieved the planned 120 miles per hour (193 kilometers per hour) in December 1932. This was just as well, since in October an Austin Seven–based Special had achieved 119.39 miles per hour (192.14 kilometers per hour).

More work in 1933 saw the body altered with an enclosed cockpit by October that year. Bert Denly, Eyston's mechanic, took the wheel of EX127. Denly was much smaller than his boss, so better able to fit in the new, slimmer bodyshell. Under his guidance, the car retook records that had been captured by Austin in the meantime.

The Magic Midget was sold by Kimber to German racer Bobby Kohlrausch in May 1934, apparently with no reference to Eyston. With Kohlrausch at the wheel, the 750 cc records were raised to 130.48 miles per hour (209.99 kilometers per hour) over the Flying Mile in May 1935 and a remarkable 140.6 miles per hour (226.27 kilometers per hour) on October 10, 1936. It was said that Mercedes-Benz took over the car for detailed study to establish how the car had achieved 146 brake horsepower and 7,500 rpm from the Zoller-blown supercharger at 39 pounds (5.4 kilograms) of boost, all from just 746 cc. These statistics would be impressive even today.

EXPANDING THE RANGE: D-TYPE MIDGET, F-TYPE MAGNA, AND THE MARK II C-TYPE

Returning to the production car situation at the start of the 1930s, Cecil Kimber wanted to expand the smaller MG range and the new Charles-designed chassis concept, taking it gradually upmarket to partially fill the gap between the two-seater Midget and the older 2.5-liter Sixes. He achieved this in two steps.

There was a four-seater Midget (the D-Type) with the same engine as the M-Type but a longer chassis; this placed any unfortunate rear passengers aft of the back axle. The second, more promising model was the new F-Type Magna, which benefited from a six-cylinder engine using the same basic components as the 847 cc four. This new 1,271 cc Six was largely the same unit as the Wolseley Hornet, but Kimber arranged to have sheet-steel pieces fixed to the engine block to disguise its Wolseley heritage. The chassis was also ten inches (twenty-five centimeters) longer—something of a mixed blessing.

These new offerings could not help but show the M-Type as an aging product of the first generation of Midgets. Sales in a period that coincided with the onset of the Great Depression were somewhat lukewarm. The C-Type came in for further attention during this period: The *Motor* of May 31, 1932, brought word of an updated Montlhéry Midget, and a week later the magazine related engine developments that justified the new version's Mark III designation. A flat-slab fuel tank was fitted in place of the boat tail of the previous Mark II, while the nose cowl was dispensed with in favor of the normal MG radiator.

For the Mark II's version of the 746 cc engine, H. N. Charles had developed an all-new cross-flow cylinder head, with four separate exhaust ports on the near side and equivalent inlet ports on the opposite side. The inlet and

EX127—better known as the "Magic Midget"—was an all-new record breaker built with the aid of MG's growing expertise. The basic shape was attributed to MG's own Reg Jackson, who built a model mock-up in his own time at home. The car is seen here at Montlhéry. *Author archive*

exhaust valves were inclined toward each other, allowing shorter and lighter rockers, and the size of the spark plugs was reduced from an 18mm to a 14mm diameter. In MG's internal tally of its EX vehicles, known as the EX Register, this AB-type engine was listed as EX130 (see chapter 8 for more details on the list). Jack Daniels was directly involved, visiting Wolseley engines to liaise with them concerning Abingdon's requirements and in the process establishing contact with Wolseley's A. V. Oak, a name that would become important in later MG developments.

Daniels told the author of a memorable trip during this engine's development:

> *Having just drawn up the initial Midget eight-port cylinder head, I was instructed to take the drawing to Stirling Metals in Birmingham. Freddie Kindell, driving a car fitted with an extra ball bearing on the forward crankshaft extension—that is an outboard third bearing—had instructions to test the engine to failure, and take me up to Birmingham in the process. We made it in one and a quarter hours, door to door, and if you happen to know the twisty road via Warwick, you will recognize a real feat. As a passenger I have certainly never forgotten it. And then, exactly a week later, we did that journey again, and I collected a specimen of the new cylinder head—still warm from the foundry—which indicates the speed and efficiency available from our keen suppliers!*

RIGHT: The Magic Midget caught the public's imagination, as exemplified by these contemporary cigarette lighters. From the collection of Michael Ellman-Brown. *James Mann*

128.62 m.p.h.
MG
[By courtesy of "The Motor."
Capt. G. E. T. EYSTON'S
"MAGIC MIDGET"
driven by A. W. DENLY at Montlhéry Track, October 19th, 1933
AGAIN PRE-EMINENT!
INTERNATIONAL CLASS H RECORDS
1 Kilometre - - - 128.62 m.p.h.
1 Mile - - - - - 128.62 m.p.h.
5 Kilometres - - - 127.65 m.p.h.
5 Miles - - - - - 127.80 m.p.h.
10 Kilometres - - - 127.23 m.p.h.
10 Miles - - - - - 125.43 m.p.h.
(Subject to official confirmation)
THE WORLD'S FASTEST 750 c.c. CAR
Issued by the Publicity Department of the M.G. Car Company, Limited, Abingdon-on-Thames.

BELOW: As ever, Kimber was quick to celebrate the latest successes of a record-breaking MG. *Aviation Archive*

ONWARD AND UPWARD: NEW MIDGETS, MAGNAS, AND MAGNETTES

By the summer of 1932, the M-Type was already considered passé, and the four-seater D-Type, while a moderate success, was not long for Abingdon production. M and D were swept away in August 1932 in favor of the new J-Type, which came in four flavors: a four-seater J1, two-seater J2 (by far the most popular), and a small Salonette saloon.

By the end of the year, there was also a supercharged version of the J2: the J3, with a 746 cc engine, and a racing model intended to take over from the C-Type. This was the new J4, which had lighter, more spartan bodywork and bigger 12-inch (30.5-cm) brakes. Only nine of the J4 were produced, which could explain why a related but unsupercharged variant, catalogued as the J5, was never actually produced for sale. By this stage, the Morris Minor had abandoned the overhead camshaft four-cylinder engine in favor of a more humdrum side-valve 8-horsepower unit.

The MG F-Type Magna had elegantly low, long lines with a flat bonnet surface; this design predated the cowled shapes that later became a hallmark of MG sports cars. This is the four-seater F1 launched as the 12/70 Magna Light Six. The engine was a sweet six-cylinder Wolseley-sourced unit. *Author archive*

At the same time, the F-Type Magna benefited from brake upgrades and some engine improvements aimed at better cooling. A J2 fitted with the six-cylinder engine became the new F2, and a new four-seater was christened the F3. The F-Type range enjoyed fair popularity and was even found with around a dozen variants from various coachbuilders, including a closed Salonette with an attractive sliding glass sunroof.

Not content with his range of Midgets and related six-cylinder Magnas, Kimber also introduced a third new range for the 1932 Olympia show. This was the Magnette, featuring six cylinders like the Magna but with a brand-new, smaller 1,086 cc unit (retaining the earlier 57mm bore but a new 71mm stroke), again derived from the four-cylinder, with much of the work undertaken by H. N. Charles and his team using project code EX131. At this juncture, Kimber embarked on a complex proliferation of both engine and model variants; the former would appear in KA, KB, K3 (racing), and eventually a larger-capacity KD version.

Body types meanwhile were offered as the four-seater K1 on a nine-foot (three-meter) chassis (including a pillarless saloon version, later the KN) and the two-seater K2 (in essence a widened J2 body but with the new six-cylinder engine), while K3 was reserved for a bespoke racing car and engine combination revealed in the spring of 1933. The K1 four-seater chassis and K2 two-seaters generally came with the KB engine.

As if this was not complex enough, MG added to the mix by replacing the remaining F-Types with the new L-Type Magnas in March 1933, but they employed the newer KC 1,086 cc six. A pillarless saloon, the KN, was added—this outlasted most of the rest of the K-series range. This

TOP: Small, compact, and frequently dressed in British Racing Green, this tiny but potent MG sports car was one of the motoring revelations of the 1930s. *James Mann*

BELOW: This 2017 photo shows Albert Koolma's Dutch-registered late J2 Midget with swept wings. Many of these little cars are still active throughout Europe. *Cathelijne Spoelstra*

confusing proliferation of models was hardly conducive to efficient manufacture, marketing, or sales, and there would be changes halfway through the decade.

For the time being, MG was at the top of its game, with motorsports exploits reinforcing the sense of pedigree.

THE MAGNIFICENT MILLE MIGLIA: DEBUT OF THE K3 MAGNETTE

We have already seen that, with the failure of the 18/100, Kimber had been disappointed in his ambitions for an almost bespoke six-cylinder MG suitable for racing by gentlemen. The successes of the very special Montlhéry Midgets and the record-breaking Magic Midgets brought MG some of the celebrity that Kimber sought, but there was another peak to climb: the international road race. The tool for this was to be an offshoot of the new K Magnette range, in the form of the K3—developed under project code EX132 from New Year's 1933.

Two prototypes were built. One was the JB1046, which was taken to the Monte Carlo and broke the record for the Monte des Mules Hill Climb. The other, built in part by Syd Enever (according to his own testimony) and registered as JB1269, was sent off to Italy the same month, via the Cornish port of Fowey, and thence to Genoa in a small boat with a cargo of fish and China clay, for an extended test session on the Continent. The objective

Cathelijne Spoelstra owns this MG J1 Midget, photographed at the Brooklands Circuit in 2022. *Cathelijne Spoelstra*

ABOVE: The three Mille Miglia K3 Magnette team cars lined up for the cameras. *From the left*: Count Lurani at the wheel of JB1475; Sir Henry Ralph Stanley "Tim" Birkin, 3rd Baronet, in JB1474; and JB1472 with Reg "Jacko" Jackson standing behind. *Jackson Family Archives*

RIGHT: The 1932 MG F2 Magna. *Author archive*

of this trip, with the second prototype under the care of Enever's colleague Reg Jackson, was to prepare the car for a team entry planned for the 1933 Mille Miglia. The team was entered by Earl Howe, whose ex-Caracciola Mercedes accompanied the MG and doubled up as tender.

During the trip, the group called in at Scuderia Ferrari, meeting Enzo Ferrari and the great Italian racing driver Tazio Nuvolari. It has often been written that they also visited the Bugatti factory at Molsheim, France, where Monsieur Bugatti was said to have studied the car and, according to John Thornley, "he declared that the front axle wasn't strong enough. . . . Jackson telephoned back to Abingdon and as a result, the design was modified in time for the race." However, it is worth noting that Reg Jackson, who was there at the time, claimed to MG historian Mike Allison that this was not true; he said that whilst Lord Howe called on Bugatti, the MG never went there; Allison suggests that it was one of the drivers, Tim Birkin, who identified the axle weakness—but presumably a Bugatti association made for a better story.

The race itself, on a figure eight of mixed roads centered on Brescia, took place on April 8 and 9, 1933. There were three British Racing Green K3 Magnettes, the only British cars competing—indeed, of the ninety-eight cars in the race, only five were not Italian. The crews were Earl Howe and University Motors salesman Hugh Hamilton in car K3001, Sir Henry "Tim" Birkin and his friend Bernard Rubin in car K3002, and Captain George Eyston and the Anglophile Italian Count Giovanni "Johnny" Lurani Cernuschi in car K3003.

Despite the inevitable patriotic fervor that surrounded the Alfa Romeos and Fiats, the Italian spectators were nevertheless intrigued by the presence of the trio of "*auto da corsa inglesi*." Birkin drove furiously—the plan was to draw out the Maseratis—and in the process he broke the Brescia-to-Bologna record at 89 miles per hour (143 kilometers per hour). Ultimately, his engine suffered and he was forced to retire.

The overall race winner was local hero Tazio Nuvolari in an Alfa Romeo, but meanwhile the other two MGs won their class and the team prize: the Gran Premio Brescia. Such was the interest in this British success that the team was granted audiences with the King of Italy and Prime Minister Benito Mussolini, the infamous *Duce*. Nuvolari later traveled to England and drive a K3 in another memorable race for the model.

MIRACULOUS RACING MIDGETS: J4 AND Q-TYPE

Following on the K3's exploits, work also continued on developing the Midget race cars, using the basic building blocks of the J2 but with larger brakes, the latest supercharged 750 cc engine, and other enhancements derived from the C-Type. MG announced the J4 in March 1933 at a price of £445 with a normal gearbox or £480 with

Artistic license makes the MG J2 Midget in this showroom poster look longer and lower than it was in reality. *James Mann*

TOP: A magnificent painting by Bryan de Grineau of the 1933 MG K4 Magnette driven by Captain George Eyston and Count "Johnny" Lurani in the 1933 Mille Miglia. *James Mann*

ABOVE: One of the Mille Miglia K3 Magnettes. This one is K3015, a 1934 car first owned by John Henry Tomson Smith; it was a costly twenty-first birthday acquisition at the time. Today, this car is known as K4015-2 as a consequence of its complicated life story. *James Mann*

the preselector unit (and £35 for the optional detachable streamlined tail). At the same time, MG also promised a cheaper unsupercharged version, the J5, for £50 less. None of the latter were sold and only nine J4s were built.

The next step in racing Midget development followed. Just as the J2 road car would be superseded in 1934 by the P-Type, so the J4 would give way to the Q-Type; even rarer than its predecessor, only eight of these were built. The tiny four-cylinder engine now produced over 100 brake horsepower at 7,300 rpm, boosted by a Zoller supercharger. The chassis was eight inches (twenty centimeters) longer and wider by three, but the power was hard to tame, prompting H. N. Charles and his colleagues to explore the newest suspension and chassis technology. This moved the dial even further from Morris orthodoxy. The story of this most exotic Midget is covered later in this chapter.

MAGIC MAGNETTES

Kimber was quick to capitalize on the success at Mille Miglia. A series of publicity exercises included press articles, luncheons, and even a book by Barré Lyndon, a flamboyant

The J4 was the racing version of the new J-Type family car. *Enever Family Archive*

author better known as a playwright. Racing successes accumulated, albeit with some setbacks that offered some important design lessons. Perhaps the most noteworthy success was when Tazio Nuvolari accepted an invitation to drive a K3 at the 1934 Ulster Grand Prix Tourist Trophy race in September. Alec Hounslow served as Nuvolari's riding mechanic. While Nuvolari spoke little English and Hounslow no Italian, they came to understand each other in the heat of the race.

Nuvolari was initially nonplussed by the K3's preselector gearbox, but he soon got the hang of it and wore out a complete set of tires within the first eight practice laps. During the race itself, Nuvolari drove his K3, No. 17, with furious skill and concentration, although Hugh "Hammy" Hamilton worked wonders with his J4 Midget, No. 25, and might have won had it not been for a bungled pit stop. Both finished, with Nuvolari forty seconds ahead of Hamilton.

It was a stunning race for both the Midget and Magnette, and an event Alec Hounslow would certainly never forget. Soon afterward at Brooklands, a single-seater K3 (chassis K3006) driven by E. R. "Eddie" Hall in the 500-mile (805-kilometer) handicap race notched up another outright victory.

TRIPLE M EVOLUTION: P-TYPE AND N-TYPE

Fired with enthusiasm in the wake of their 1933 racing successes, the small design team turned their attention later in the year to upgrades for their core models. The racing K3 Magnette had life in it yet, so it was overhauled for the 1934 season, but the new N-Type models soon usurped the remainder of the K-Type family. The roadgoing Midgets segued toward the new P-Type, although before that the J2 was face-lifted for the 1933 Motor Show, with elegant flowing wings in place of the earlier cycle type. The J1 four-seater was dropped, as was the old-hat supercharged J3. The P-Type arrived on March 2, 1934, along with the Q-Type racing version described above.

The end of March saw the announcement of the N-Type Magnette range, with a 1,271 cc version of the KD six-cylinder engine as seen in the earlier K2 models. The gearbox in the

The Q-Type Midget was a lightweight racing car with prodigious power. In fact, the chassis proved barely capable of handling the output even in the hands of capable drivers. *Enever Family Archive*

N-Type was derived from a Wolseley unit, and there was a Bishop cam type of steering unit in place of the Marles-Weller type that had been previously used. Despite the arrival of the new N-Types, the L-Type Magna remained available as the Continental Coupé, with the older 1,087 cc engine and priced at £350, £15 more than the new N-Type four-seater.

The longer-chassis KN Magnette was also offered as the £399 Pillarless saloon, an evolution of the pillarless K1 of 1932. The KN was soon described by one *Motor Sport* writer as "hot stuff" in light of its name being an obvious homonym for cayenne pepper. Its engine had 57 brake horsepower at its disposal (eighteen more than the original K1), so this was a fair appellation. The K3's racing successor, the NE Magnette, appeared later in 1934. The Tourist Trophy race on September 1 saw six cars entered. Impressively, Charlie Dodson brought his NE home just seconds ahead of E. R. Hall's Bentley, at a speed just 3 miles per hour (5 kilometers per hour) slower than Nuvolari had achieved the previous year in his K3.

Both the Midgets and the Magnettes as a whole benefited from slightly longer wheelbases, while the P-Type saw a new version of the successful 847 cc four-cylinder engine, now with a stronger three-bearing crankshaft, upgraded cylinder head, and four-speed gearbox. The styling set the tone for MG sports cars for the next twenty years, with a double-humped scuttle, cutaway door tops suited for resting elbows, and a short tail with upright fuel tank and externally mounted spare wheel.

June 1935 saw Kimber supporting an entry of P-Types at Le Mans, with the novelty of an all-female driving crew, under the management of George Eyston. The six drivers were Barbara Skinner, Margaret Allan, Doreen Evans, Colleen (Mrs. Hugh) Eaton, Joan Richmond, and Margaret (Mrs. Gordon) Simpson. They were soon dubbed "Eyston's Dancing Daughters." Because the goal was to finish rather than strive for class victory, all three cars finished some way down the field—but so secured the desired publicity.

Hypothetically, the rationalization that accompanied the replacement of the J and K ranges with their P and N counterparts could have generated better economies of scale and bolstered MG's wavering profits, which would be up one year and down the next. Production in 1932 peaked at 2,400 units, but had declined steadily in 1933 and 1934 to 2,100. It plummeted to less than 1,300 in 1935, a year of heavy financial losses for MG.

Lord Nuffield had allowed for the fluctuations because he put his faith in Kimber, and he was far less hostile to motor racing than some have suggested (he often gave the prize money for races). Still, the fact that Kimber was spending more on motorsports and specialized components hardly helped when the time came to further rationalize his business interests in 1935.

ABINGDON'S FINEST: THE REVOLUTIONARY R-TYPE

The Q-Type combined a fantastic powertrain with a chassis at the limits of its practical development. Changes to the rest of the structure would be necessary to exploit the further potential of the supercharged 750 cc miracle. The first that the motoring enthusiast world knew of the planned updates came in the March 1935 *MG Magazine*, where preliminary details appeared for two racing Specials with single-seater bodies. In the article, Kimber called the new cars the "MG Monoposto Midget, 750 cc" (the new R-Type) and the "MG Monoposto Magnette, 1,100 cc" (planned, but never produced; possibly an S-Type). Elsewhere Kimber had written to the motoring press seeking popular support for changes to the racing formulas to benefit these new cars.

The R-Type was a true tour de force, MG's proposed entry into the field of truly modern, state-of-the-art motoring exotica. Other than the engine, which was essentially a Q-Type in form, the chassis, body, and most of the other fixtures and fittings were about as far removed from a Morris 8 as one might imagine. Jack Daniels worked on drawing the car in late 1934 into early 1935, mostly to half-scale on a vertical board. Recent arrival was Bill Renwick, a bright mathematician who worked on the theory behind an all-independent suspension. Daniels recalled that "there were parallel forged wishbones for the front suspension, and the top ones—at three-eighths of an inch [.95 cm] diameter—looked like little more than a couple of knitting needles; the drivers saw them and said no way were they were going to drive the car with such thin wishbone arms, and so we made them thicker just to keep the drivers happy!"

Lord Nuffield lays a paternal hand on the K3 Magnette in which celebrated driver Tazio Nuvolari won the 1934 TT race. *Author archive*

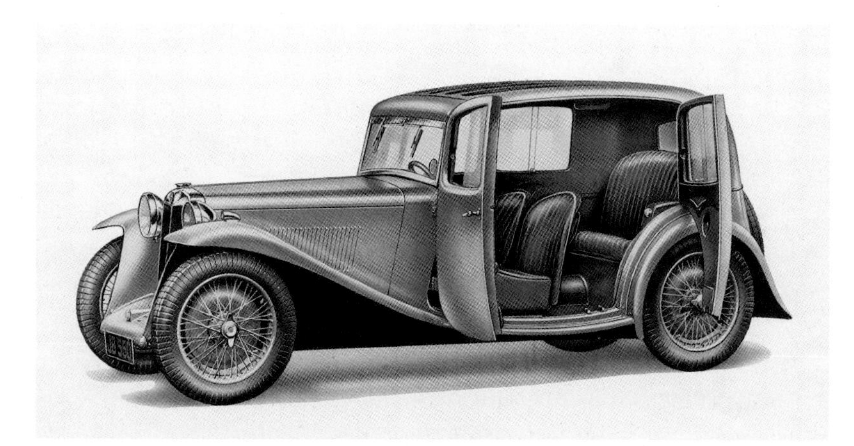

H. N. Charles determined that, to make the new suspension system work, its underlying structure needed to be as rigid as possible. This meant eliminating all of the chassis flex seen on most contemporary sports cars, such as the Q-Type, an example of the old school. Charles resorted to a box-section wishbone chassis before handing off the plan to Frank Stevens, who fabricated the prototype. (Stevens had been responsible for modifying the chassis of Old Number One ten years earlier.) With its lightweight chassis and minimal bodywork, the resulting R-Type tipped the scales at a featherweight 1,421 pounds. That chassis layout and form would next be seen on the 1963 Lotus Elan—MG's R-Type was truly ahead of its time.

Kimber and Charles knew they were pushing into new design territory, and they worried about the possibility of infringing patents with the new suspension. To avoid this, they employed Eric Walford, a patent agent, to study existing patents. The one that caused the most concern was Roesch & Clement Talbot's patent for an independent suspension system, but in the end Walford concluded that Abingdon had no need to worry: After the war, MG's research into these patents offered useful support for the Morris Minor, a collaborative effort by Jack Daniels and Alec Issigonis.

The R-Type first raced at Brooklands on May 6, 1935, at the Junior Car Club International Trophy Race, with a team of four entrants. The independent suspension was a revelation, but the cars exhibited a peculiar body roll. It was an issue that could have been eliminated with further work, but there was no time to pursue a solution. This left the drivers struggling to adapt to the altered suspension and inflexible chassis.

Bill Everitt and Malcolm Campbell drove RA0260 to a well-deserved class win, with Doreen Evans immediately behind in RA0255.

Later that month the four-car R-Type team appeared at the Isle of Man Mannin Beg race. Broken driveshafts, partly a consequence of overheated hydraulic shock absorbers, meant that none finished. Back at Abingdon, Charles quickly began to engineer improvements. Events outside of design fixes precluded this work, which was never completed.

HIGH AMBITION: AN ALL-INDEPENDENT-SUSPENSION FUTURE

The R-Type was supposed to be just the start of Kimber's ambitious plans. In addition to the related all-independent Magnette mentioned above, there were tentative plans for a larger roadgoing saloon with sophisticated suspension built along similar lines. Records in the EX Register reference EX150 as a "3½ Litre Independent Car." The implication was obvious: Kimber was once again gazing jealously at the Bentleys of similar engine size, although within the Nuffield family his engine choice would have been limited to the heavy six from the Wolseley Super Six, which had evolved from a lorry engine.

Over the years, rumors circulated that Kimber was looking outside the Morris-Wolseley family for a possible engine. Candidates included a bespoke V-8 or a Blackburne straight six. Certainly work was undertaken by Mulliner for some sleek upmarket coachwork.

Jack Daniels told the author that he had drawn up the chassis, but he couldn't recall the engine detail. "We built one," he said, "and it was scheduled for the 1935 Motor Show at Olympia. The car was built on the lines of the R-Type, but to overcome the problem of the roll angle, Bill Renwick sat down and designed a preloaded anti-roll bar—which I drew up for him—which had a torsion bar concentric with a torsion tube, with one hundred pounds preloading, so that when the car went into roll mode, there was already a force restraining the roll

OPPOSITE TOP: An MG P-Type with the sleek "Airline" coupe body style, designed by Henry William Allingham. *Author archive*

OPPOSITE MIDDLE TOP: A Berkshire-registered 1934 MG P-Type Midget. *Author archive*

OPPOSITE MIDDLE BOTTOM: The MG KN Pillarless Saloon was an elegant sporting saloon in its day. "Trafficator" signaling devices emerged from the rear part of the roof (there being nowhere else to mount them). *Author archive*

OPPOSITE BOTTOM: The MG NA Magnette four-seater tourer. *Author archive*

BELOW: GeorgeEyston in Brooklands trying out a new GBY913. *Alamy*

RIGHT: Leonard Lord was about as diametrically opposed to the urbane and sophisticated Cecil Kimber as one could imagine. Variously brilliant, mercurial, and purportedly saddled with an inferiority complex, Lord shook up Abingdon in the summer of 1935. For all that, his actions helped reverse the decline in MG profitability. *Author archive*

BELOW: For the 1935 24 Hours of Le Mans, Captain Eyston entered a team of three specially prepared P-Type Midgets driven by an all-female team. The press dubbed them "Eyston's Dancing Daughters." *Alamy*

action." Daniels also confirmed that the car would have benefited from a "classy, typically MG look." But in the event, it came to naught.

MG AT THE CROSSROADS: THE TAKEOVER OF 1935

It has occasionally been suggested, unkindly, that William Morris allowed MG to be rolled into his main business either out of pique or from dislike of motor racing. The reality is far more complex. He was now officially Lord Nuffield, having been advanced to the peerage as Baron Nuffield, of Nuffield in the County of Oxford, on January 15, 1934. On one of his increasingly frequent overseas trips in 1934, he was introduced to a group as the maker of Morris cars. When that drew a blank, it was added that he was also the owner of MG, which drew smiles of recognition. For the great industrialist, although inherently a modest man, this may have been somewhat jarring.

Other events at the time worked to cool any enthusiasm Lord Nuffield may have had for racing. Driver Kaye Don and his mechanic, twenty-eight-year-old Francis Leonard Cyril "Frankie" Tayler, were involved in a collision on May 29, 1934, during an unofficial late-night practice run for the Mannin Beg race on the Isle of Man. The accident took place on the Laxey Road north of the city of Douglas, and Tayler succumbed to his injuries at Noble's Hospital. The popular young MG mechanic had started at Morris Garages in 1923. The subsequent court case against Don, which resulted in a custodial sentence, led to ironic scoffing from the press of MG's jealously guarded standard of "Safety Fast." The media coverage was equally unkind to the benevolent image of Lord Nuffield, whose philanthropy in the field of medicine was widely known. Tayler's death also shook Lord Nuffield personally: H. N. Charles was convinced that it turned Nuffield against motor racing.

A more fundamental challenge, from a financial perspective, came from two court cases associated with the super tax, which was aimed at those with a high net worth. Lord Nuffield had successfully fought the tax thus far, but the cases nevertheless left in their wake a lingering concern about future tax liability: Ownership of a disjointed web of business interests would remain a red flag for the tax authorities. In the midst of this situation, Lord Nuffield placed great trust in a bright engineer, Leonard Percival "Len" Lord, to organize Cowley with greater efficiency.

By the mid-1930s, market conditions had recovered from the recession that had led off the decade, so it was decided that Lord Nuffield should sell most of his personally

The 1935 MG NB-Type Magnette. *Author archive*

owned businesses, like MG and Wolseley, to Morris Motors. He retained the Morris Garages and Wolseley Aero Engines). Len Lord argued persuasively that there was little logic in the privately owned Wolseley selling a significant proportion of its component output to the separate Morris Motors business.

So it was that, on July 1, 1935, the MG Car Company was sold to Morris Motors Ltd. and Len Lord became the company's managing director. Kimber retained his directorship and at the same time became a director of Morris Motors Ltd., thus rising in stature and remuneration. This arrangement lasted for only a year, though: Len Lord quit the company, eventually accepting a position as works director at Austin Motor Company's Longbridge factory in the English Midlands, and Kimber was reappointed as managing director of the MG Car Company on August 24, 1936.

It was not all plain sailing, however: Len Lord took a dim view of MG's almost sinusoidal profit lines and decreed that there should be significant efficiency savings. A loss of £30,667 (including tax) by the MG Car Company for the first eight months of 1935 certainly did not help Kimber's case. As historian Peter Seymour notes, "It is a sobering thought to consider that if W. R. Morris had not been a wealthy man, the MG Car Company Limited would probably have become bankrupt in 1935 and disappeared along with many other motor manufacturers of that period." In context, all of Lord Nuffield's companies nevertheless made a profit of £1.44 million in the same year (equivalent to around £81 million in 2023).

Accordingly, in his position as managing director, Len Lord closed the racing shop and moved the core design function to Cowley, where he was determined that future MGs would hew more closely to Morris orthodoxy, abandoning dalliances with unique engines or fancy suspension systems. Lord was quoted as saying that, in future, MGs should be simply "tarted up Morrises and Wolseleys." The *Autocar* wasted no time in reporting that "the decision that the MG company is to cease racing forthwith has come as quite a shock to a lot of people," going on to suggest that the cessation of racing would have an inevitable knock-on effect upon factory support for private racing of MG sports cars. It is worth noting that the finances for the last four months of 1936 showed a profit of £16,037, and that the 1937 profit of £25,436 was the highest ever recorded up to that time.

Design work now shifted to Cowley. H. N. Charles, Jack Daniels, and colleagues George Cooper, Eric Selway, Geoff White, and another named Lewis went with it, while Syd Enever stayed at Abingdon as the liaison man. Bill Renwick soon left Abingdon, and Charles

A new one-and-a-half litre
FOUR DOOR SALOON, TOURER & FOLDING HEAD FOURSOME

TOP LEFT: The MG SA Two-Litre came with a ten-foot (three-meter) wheelbase—the largest MG to date—and an impressive curbside presence rivaling that of William Lyons's SS Jaguar, a car pitched at the same market segment. The MG used a modified Wolseley Super Six engine, while Lyons's car used a Standard Motor Company powerplant, both leaning on humbler origins in their quest for pedigree. *James Mann*

TOP MIDDLE: The next part of the Cowley-Kimber MG saloon plan was the VA—or the MG 1.5-Litre, as it was more formally known. A popular car, its commercial viability was impaired by Kimber's insistence on applying more than forty octagonal MG emblems on the car. *James Mann*

TOP RIGHT: The 2.6-liter WA MG was the last gasp for the bigger MG saloons. It proved short-lived thanks to the outbreak of World War II. After the conflict was over, the company decided there was no longer a need for such a large and stately MG. *James Mann*

was drawn briefly into other work. At Cowley, the technical director was Robert Boyle, supervising chief designer Bill Seddon. Soon after, the young Alec Issigonis arrived to work on suspension, with Tom Brown on engines and Roger Frederick Cordey on axles.

"For about a year, that was the set-up, and then Boyle was sacked," Daniels told the author. "Tom Brown went back to Morris Engines, Cordey disappeared, and in 1937 came A. V. Oak, the new chief engineer, brought in from Wolseley." As technical director, Vic Oak was given overall responsibility for Morris, Wolseley, and MG cars.

At this juncture, MG design work was split, but the sports car work was overseen by a Jack

Grimes as "Section Leader for MG" (Daniels claimed that, in his words, Grimes did "bugger all") and delegated to Jack Daniels and George Cooper. As well as finishing off the last of the P- and N-Types (see below), their key work was on what would become the new T-Type sports car.

ENDGAME FOR THE MMM CARS: PB AND NB

Although design authority and work shifted to Cowley, MG production remained at Abingdon (although, as we shall see later, senior management sometimes had other ideas on that score). This meant that, while design for the new S- and T-Types was underway, the existing model ranges had to be sustained and refreshed. The new PB Midget, with a larger-bore 939 cc four-cylinder engine, retailed at £222 and initially supplemented the earlier P-Type, now known retrospectively as the PA at a price of £199 10s. The radiator grille sported painted vertical slats similar to those of the late lamented 18/80. The dashboard featured a novel 30-mile-per-hour (48-kilometer-per-hour) warning light, illuminating when the car exceeded speeds 20 miles per hour (32 kilometers per hour); this alerted the motorist when driving through built-up areas where the new speed limit was in force. The PA proved a difficult sell when set beside the newer PB, and in late 1935 the final twenty-seven PAs were reworked into PBs at the factory.

As with the Midget, the N-Type Magnette was given a face-lift, including a lower-profile body line and grille slats similar to those of the PB. The doors were now hinged from the front rather than the rear, with distinctive long-plated, spearlike hinges. Internally, the speedometer and tachometer were separated, the instrument in the earlier model having been combined. The NB proved to be a desirable sports car, but it would soon be overshadowed by much bigger six-cylinder MG, described below. In the autumn of 1936, the NB was discontinued and, with its passing, the classic MG overhead cam six was no more.

THE COWLEY MGs: S, T, V, AND W

As we have seen, the P-Type Midget survived for some time following the events of summer 1935, but more immediate urgency was afforded at the upper end of the range. Rumors circulated around Motor Show time of a forthcoming MG two-liter and, indeed, a large, distinguished, but heavy saloon was under development with a Wolseley-based overhead valve engine. Initially 2,062 cc, it was soon expanded to 2,288 cc to provide enough power to counteract the weight factor (it soon grew again to 2,322 cc). The wheelbase was 10¼ feet—unprecedented for MG. Unfortunately for MG, development delays allowed William Lyons to steal a sales advantage with his low and stylish 2.5-liter SS, also shown at the 1935 Olympia Motor Show.

OPPOSITE BOTTOM: The MG T-Series prototype. The 1936 TA was followed by the short-lived TB of May 1939, which looked very similar but offered mechanical refinements, most notably the new 1,250 cc XPAG engine in place of the earlier 1,292 cc MPJG unit. *Author archive*

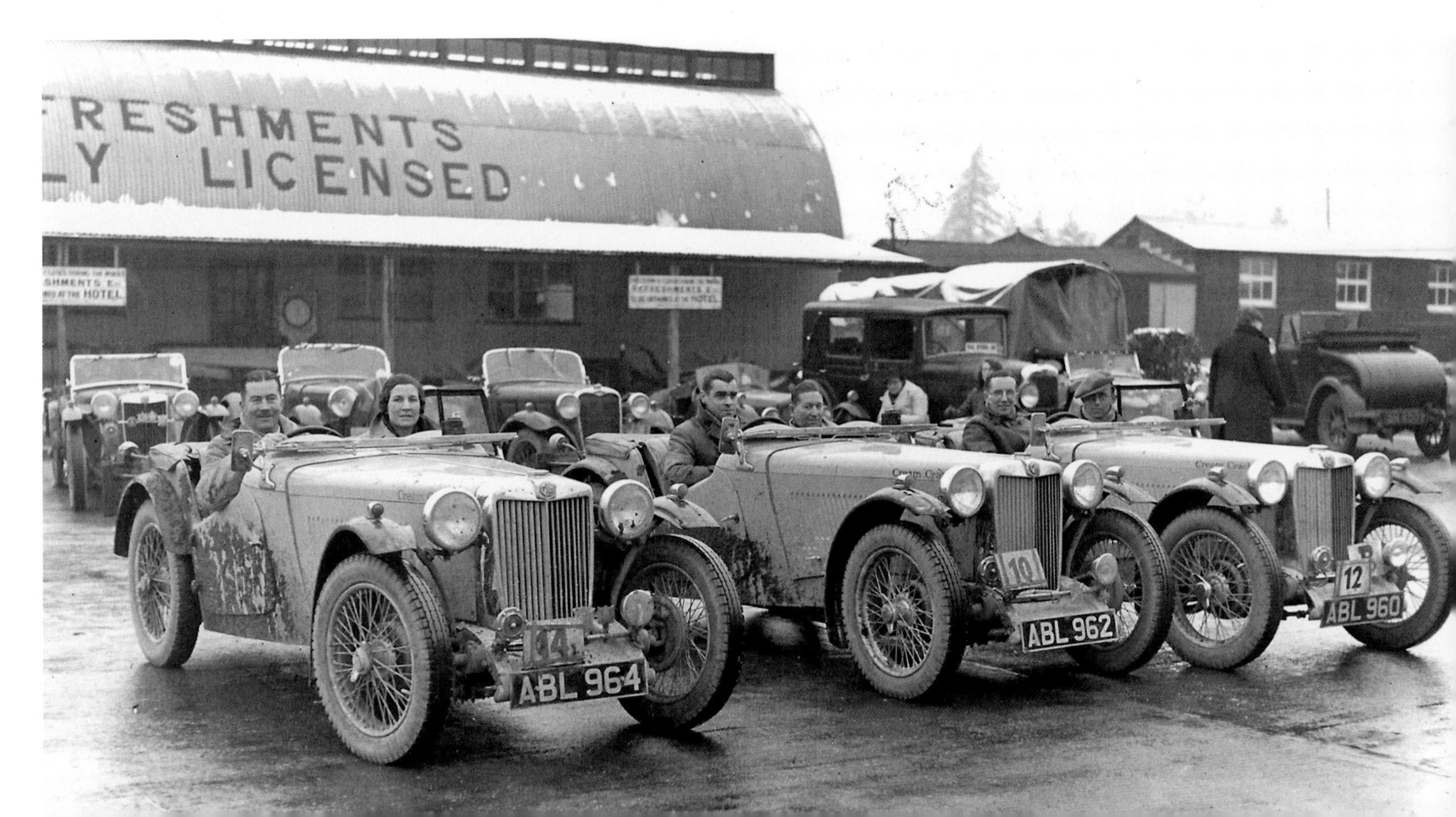

BELOW: The MG Cream Cracker Team in March 1937, parked at the end of the trial. All three TA Midgets had 1,292 cc engines. They were driven by J. E. S. Jones, H. K. Crawford, and L. Welch. *James Mann*

A. B. Langley behind the wheel of one of the MG Musketeer TAs in the Sutton Coldfield and North Birmingham AC's Colmore Trophy Trial (Colmore was a major Nuffield Group depot). *Author archive*

Next in the new lineup was the new Cowley-designed MG Midget, announced in June 1936 as the T-Series, which owed much of its design to Jack Daniels and H. N. Charles. They managed to combine much of the experience they had gained during their time at Abingdon but on a chassis that in general proportions was closer to a Magnette than the petite P-Type Midgets. Under the bonnet, pride was dented with the adoption of a long-stroke 1,292 cc Wolseley 10/40 engine with overhead valve gear. The first car tried by the *Autocar* was criticized by the magazine, which complained that "it was odd to be without the familiar exhaust burble, for there was no real sound to the exhaust pipe." Pretty quickly, the production cars were improved in this regard.

As these designs were being implemented, Charles and Issigonis both sought to develop an advanced suspension design. According to Daniels, they did not always see eye to eye, and Charles was the first to blink. "Charles resigned and went to Rotol where he redesigned the variable pitch air-screw."

In 1937, Daniels moved over to work with Issigonis, leaving George Cooper working on MG design for a while. Daniels said that Vic Oak teamed him up with Issigonis to give the latter an anchor in the real world: "I had met up with A. V. Oak in the past, when he had been chief engineer at Wolseley: I used to go there on mechanical objectives, mainly pertaining to the Midget—including of course the crossflow cylinder head for the Montlhéry Midget—and so he was aware of my potential. Oak could see that Issigonis had a gift, but needed someone experienced in practical issues, and I was his choice." Other than working on a new front suspension, Issigonis seems to have had virtually no involvement in the design of prewar MGs.

Third in the letter sequence of these Cowley-designed cars was the VA, with another Wolseley four-cylinder engine, of 1,548 cc. There was much potential for a 1.5-liter saloon of this kind, with its nine-foot (three-meter) wheelbase, not least because Williams Lyons was competing in the same market as the S- and V-Type classes with his own rakish SS models. Unfortunately, from a cost perspective, Kimber was obsessed with octagons. As John Thornley later recalled, "He put no less than forty-seven onto the VA; special handles, instruments, horn pushes, etc., and so the cost of the VA went up until it was no match for the SS car."

Kimber was perhaps unaware of this criticism. In 1944 he wrote a paper on "The Trend of Aesthetic Design in Motorcars," where he stated of the VA: "This car was designed around the seating accommodation for four persons, and the first drawing was produced in September 1935. No less than four sample bodies were built before the design was finalized and it follows this author's ideas on design as set out briefly in this paper." Kimber ran a VA as his personal car, but with the engine bored out to give a capacity of 1,708 cc.

In 1938–1939, around the time of the Munich Crisis, these MG models moved into what would become their final prewar guise. The T-Series became the TB, with a new, more sporting 1,250 cc XPAG engine, designed by Claude Bailey of Morris Motors Engines Branch and derived from that of the latest Morris Ten. Bailey is better known for his later work with Jaguar on the XK and later great engines.

Production of the MG TB Midget began in May 1939, but for obvious reasons stopped in the autumn, with just 379 examples built. There were also efforts to up-engine the SA in the wake of growing SS 2½-Litre sales. The EX Register records EX160 as a "3.5-litre car designed at Cowley," which may have comprised use of the bulky Wolseley Super Six 25-horsepower unit. The next entry, EX161, became the 2.6-liter upgrade from the SA to create the WA, with a 2,561 cc six-cylinder engine. There were a number of enhancements to both size and style, and like the SA the new vehicle was offered with open as well as closed body styles. Like the TB, production was short-lived, but unlike the Midget there would be no postwar renaissance for the big MG sixes.

TRIALS TRAILBLAZING

Lord Nuffield was now wary of spending shareholders' money on circuit racing, perhaps more than he had been when the risk was to his own pocket. He could nevertheless appreciate the benefits of some forms of motorsports to the image of MG sports car sales, especially if the day-to-day costs were mostly borne by a third party. Accordingly, two areas of MG motorsports activity thrived in the latter half of the 1930s: trials events (where Old Number One had become famous) and the rarefied field of record breaking.

Trials events were exactly what the term implied: driver tests of moderately improved road cars—usually driven by the car owners—in testing circumstances such as hill climbs and rugged off-road tracks. The beauty of these trials was that spectators could imagine

J. M. Toulmin drives his 1,708 cc Cream Cracker TA Midget at the Exmoor Experts Trial. He won his class for cars of up to 2.5 liters. *Author archive*

See it has "TripleX" all round—every MG has
The Sports Car
6D
THE M.G. OWNERS' MAGAZINE

The EX135 unveiling was widely reported at the time. Here, it features on the cover of the *Sports Car,* the in-house MG publication of the day. Kimber is seen at left looking toward the camera. *Author archive*

buying similar cars from their local dealer. In addition, there tended to be a fairly level playing field, and the events themselves were generally accessible, whether by car or motorcycle. They were far less the preserve of the upper echelons of society, who tended to frequent the top motor racing circuits. Lord Nuffield was quite happy to fund trophies for trials events, certain in the knowledge that his corporate balance sheets would not hemorrhage cash in the process.

MG entries in trials began in the 1920s. With the closure of the factory racing team in 1935, though, attention focused on the colorful trials teams. These included the Cream Crackers—initially P-Types run by J. Maurice Toulmin, R. A. "Mac" Macdermid, and Jack A. Bastock; later T-Types run by Toulmin, Ken Crawford, and J. E. S, Jones—and the Musketeers—started off as a trio of NE Magnettes, run by Sam Nash, Lewis Welch, and Freddie Kindell; later switched to hybrid Specials and, ultimately, T-Types. (Maurice Toulmin, it should be noted, had no connection to the well-known Toulmin Motors of the postwar period.) As Jonathan Toulmin, Maurice's son, explained to the author, "Dad was the only Cream Cracker or Musketeer driver to remain on (and captain) his team throughout their entire existence. In the case of the Cream Cracker team, this was from Good Friday 1935 (April 19) until the end of the Exeter Trial on January 7, 1939."

RECORD BREAKING WITH THE MAGIC MAGNETTE

The other strand of MG motorsports provenance to survive the racing purge was that of record breaking. Captain Eyston ordered a K3 with another special offset-driveline chassis, which was longer than the normal K3 by four inches (ten centimeters), recorded as EX135. The engine was basically the 1.1-liter K3 six, but with a bronze head and a Powerplus supercharger (for which Eyston had the agency). The car came with two bodies: a narrow, track-racing and record-breaking body, and an alternative, broader, more conventional body for road-racing. The track body was finished in the MG colors of brown and cream, running as longitudinal stripes front to back.

The MG publicity dubbed it the Magic Magnette, but the popular press christened the car the "Humbug" on account of its resemblance to a well-sucked mint candy. The road-racing body was far less elegant and was known irreverently at the factory as the "Coal Scuttle." In this form, EX135 would race in the same events as the more conventional K3s under Eyston's hand—notably at the 1934 Mannin Beg race. In October 1934, to Eyston's annoyance, Kimber had a Zoller supercharger fitted for record-breaking work. The efforts at the world hour record were stymied, though, when the Zoller's casing split, which likely pleased Eyston—he later successfully chased records in Class G of up to 128.7 miles per hour (207.12 kilometers per hour), beating a recent record taken by Ronnie Horton in another K3-based car.

Soon after the news arrived that MG's racing shop was to close, efforts to break records were abandoned. Captain Eyston ceased his MG record-breaking endeavors, though twenty years later he took up the pursuit again. Stepping into the gap left by the official end of record breaking was Lieutenant-Colonel A. T. Goldie Gardner, who had successfully raced MG Midgets. He acquired the Horton MG K3 single-seater and brought it to Brooklands in August 1936. There, as John Thornley later wrote, "he managed to rekindle Kimber's enthusiasm. . . . [Kimber] agreed with Gardner to prepare the car at the factory." Clearly this needed the sanction of Lord Nuffield, whose permission Kimber sought. Lord Nuffield could see the merit in chasing glory largely at someone else's expense, and so the Magic Magnette story entered its next phase.

Gardner took the car to Frankfurt for the annual Nazi-sponsored racing event, Speed Week, in October 1937, where he took his own records for the mile and kilometer. In the process, he achieved 148.8 miles per hour (239.47 kilometers per hour)—which made reaching 150 miles per hour (241 kilometers per hour) from 1,100 cc an enticing target, perhaps making the prospect of 200 miles per hour (322 kilometers per hour) a possibility. The MG intrigued their German hosts (mostly focused on the glorious monsters from Auto Union and Mercedes) and drew the attention of Auto Union's Eberan von Eberhorst, who suggested that what MG needed was a streamlined body similar to the record-breaking Auto Union. Kimber secured Lord Nuffield's backing, and EX135 was acquired to use as a basis for the new car.

The body was designed by Reid Railton, who had also been responsible for John Cobb's world-land-speed-record car: A smooth, futuristic alloy envelope, it certainly looked at home in the company of the bigger, more powerful German record breakers. The German press and racing fraternity admired the heroic pluck, military bearing, and demeanor of "the English Major," who represented to them the admirable qualities of Britain at the time.

The car became known as the Gardner-MG Record Car, though at the MG works it was still EX135. In 1938, Gardner was back on German soil again, raising the ante when he brought his sleek-bodied projectile tantalizingly close to the 200-mile-per-hour (321-kilometer-per-hour) objective. He achieved 187.616 miles per hour (301.93 kilometers per hour) for the Flying Mile and 186.567 miles per hour (300.25 kilometers per hour) for the Flying Kilometer, with a highest one-way speed of 194.386 miles per hour (313.83 kilometers per hour). The target was obvious for 1939: to exceed 200 miles per hour (322 kilometers per hour), if possible by a generous margin. The plan was laid in conjunction with Reg Jackson, and Syd Enever of MG was to take the 1,100 cc record, quickly rebore the car onsite, and then take the 1,500 cc record immediately afterward.

In May 1939, these ambitions were realized at Dessau, with the 1,100 cc records falling on May 31 and those for the 1,500 cc class two days later. In reality, the rebore was modest, just enough to raise capacity to 1,105 cc, but sufficient to reach the next class. The best speed achieved was a one-way best of 207.37 miles per hour (333.72 kilometers per hour) in 1,100 cc guise.

Despite plans to return the following year, the outbreak of war in September put paid to further thoughts of record breaking, as it did for all car production at Abingdon.

The Gardner car returned to Germany in summer 1939 to take more records. The team laid tentative plans to return later that year, but those proposals were shelved when war erupted in Europe. *Enever Family Archive*

THE
BARLEY MOW
INN
CBL 192

3

War Years

WAR FOOTING

The declaration of war in September 1939 halted many aspects of the motoring world as it did with nearly everything else. Goldie Gardner's plans to return to Dessau were obviously out of the question, and motor-racing events were withdrawn as the circuits closed. Car production was also quickly suspended in the interest of clearing factories to accommodate war work.

This meant that the relatively new TB Midget models, and the WA saloon and tourers, were all short-lived: Only the former would have some kind of production continuity after the war. A new model, intended for the 1939 Motor Show, was put on hold. The MG Ten, as it would possibly have been known, did appear after the end of hostilities, although one of the prototypes wound up being used throughout the war for general duties.

Gardner's record car was squirreled away in a vacant clothing factory that the MG works took over at St. Helen's, at the far end of the town, where the company set up as a subsidiary factory and stores. This location saw the storage of most of the factory spares and much of the tooling that was cleared away from the MG Pavlova Works factory floor to make space for the anticipated call on the MG workforce to aid in the war effort. Although actual car making had ceased, there was a popular belief—as had been the case at the start of the last war—that the conflict would be over soon and the United Kingdom would quickly return to normal.

OPPOSITE: A publicity photo from 1940, featuring Cecil Kimber's own VA 1.5-liter saloon—a personal favorite that his family kept for years after the war. The MG TB Midget CJB 59 is the prototype that later served as the basis for the TC. *Rutger Booy, Conam Archive*

Had events gone differently in September 1939, visitors to the 1940 Earls Court Motor Show may have seen something like this car on the MG stand: a small new saloon with the latest 1,250 cc XPAG four-cylinder engine, independent front suspension, and bodywork related to the new Morris Ten. This MG Ten would have to wait until the end of World War II. *Author archive*

Like many factories, MG set out in the confident hope that war work of some kind would come their way. If the corporate masters at nearby Cowley had similar thoughts, though, they were focused more on the plants in Oxford and Birmingham, manufacturing plants that became deeply involved in making tanks, military vehicles, and aircraft. "Sleepy" Abingdon-on-Thames, where the MG factory had focused on car assembly, was all but shut down. Accordingly, the resourceful management at MG set out to look for work.

At the same time, Kimber ran a series of patriotic advertisements, graced with his signature, declaring "It reminds me of my MG," with stirring drawings of Spitfires and other war machinery. Sterner observers within the wider Nuffield Organization were happy to criticize the continuation of the "cult of Kimber" when there were, they suggested, other far more serious national issues to address.

Kimber was focused on more than publicity, however: He needed to find a place to occupy his workers with "serious work." A personal friend of his, John Howlett, had founded Wellworthy, a major precision engineering company based in Lymington, Hampshire, which manufactured piston rings. He had been appointed the Southern Area Emergency Services Chairman at the outbreak of war. To supplement the main Wellworthy plant, now busy with war work, Howlett had set up additional factories such as an operation that took over the tanning shop building at the Pavlova leather works, where some five hundred MG workers were based, a high proportion of whom were women. The idea for this mutually beneficial association came from Kimber.

In his biography, *The Guv'nor*, Howlett explained how Kimber's sister-in-law, Mrs. Hamilton, became involved. She "took a job as a sort of personnel assistant, with an office in with the Pavlova people. Her main job was the running of Fyfield Manor, a house I leased from St. John's College as a convenient stopover for Wellworthy people visiting our Abingdon works and as a center for informal business conferences." (The fourteenth-century house, more accurately "The Manor House," which just happens to be in the village of Fyfield, between Abingdon and Oxford, may easily be confused with another place called Fyfield Manor, in another location.) When Kimber and his new wife Muriel felt obliged to vacate their rented home, the Miller's House, they stayed with Howlett at this grand residence in sleepy Fyfield.

Through Howlett's excellent aviation industry connections, Kimber was able to secure a contract to support the production of the Albemarle bomber. This aircraft featured a complex nose cone and a complicated forward leg for a tricycle undercarriage. MG arranged to build these components, eventually turning out over nine hundred. For Cecil Kimber, this contract proved something of a Pyrrhic victory. Writing after the war for a factory-published booklet, "MG War Time Activities," the anonymous author (actually George Propert) stated with self-deprecating humor: "It is quite true that if we saw something in the sky we could safely say it was an aeroplane, but as for knowledge of the detailed intricacies of production, this was a closed book to us."

KIMBER'S LAST DAYS

For some within the wider Morris Motors management, the Albemarle move was a step too far. The effort, they felt, served the glory of Abingdon alone rather than placing the corporation in the best light.

To complicate matters, recent events in Kimber's personal life had strained his relationship with Lord Nuffield. He had separated from his terminally ill wife, Renee, in 1937, after having begun an affair with another woman, Muriel Dewar. According to family members, Renee's behavior had become irascible and unpredictable as her illness got worse; in hindsight one can imagine the household atmosphere became tense. Cecil Kimber married Dewar on June 25, 1938, only three months after Renee's death from cancer in April. All this was viewed with distaste by Lady Nuffield, the self-appointed guardian of company morals at Morris Motors as well as the protector of her husband's preeminent social standing.

Lord and Lady Nuffield had liked the first Mrs. Kimber, and additional interlacing relationships within the company played a role in driving a wedge between the owner and his managing director. The future deputy chairman of Morris Motors, Miles Thomas, had married the future Lord Nuffield's secretary Hylda Church in 1924, and Mrs. Thomas had become Renee's close friend.

Abingdon took on many different tasks during the war, such as rifle production. *Author archive*

The rather crude end of the Kimbers' marriage may have colored Mrs. Thomas's view of Cecil Kimber, as it is said to have done for Lady Nuffield.

Some of this is conjecture, but there seems little doubt that these matters exacerbated tensions within the management office and conspired to aid those jealous of Kimber's status to move against him. Many then and since viewed what happened next as a cynical move, perhaps tinged with spite and jealousy. It is easy for us in the twenty-first century to deem these opinions judgmental and morally outdated, but in the context of the time, divorce and extramarital affairs were scandalous. The country was also coping with the story of King Edward VIII's recent abdication in favor of marriage to the divorced Wallis Simpson.

Kimber's daughter, Jean Cook, suggested to the author that her father's relationship with Lord Beaverbrook and the friendship with John Howlett made some form of action necessary. "My father said that a directive came from Morris Motors to centralize the issuing of unemployment and insurance stamps, which would have meant sacking a faithful employee, a single woman who supported her widowed mother. My father refused, and next day Miles Thomas arrived to demand his resignation."

The coup de grâce thus came in a visit by Thomas to Abingdon in November 1941, where he suggested that Kimber should "direct his energies elsewhere." With that, Kimber's nineteen years with Morris Garages were summarily ended.

Cook describes the outrage of the Abingdon workers when they heard the news of Kimber's dismissal, going so far as planning a strike. Her father would hear none of it, insisting that they had important war work to do that should not be affected by the exit of one man. Even so, his daughter, who was fifteen at the time, saw that Kimber was quietly shattered. "The official present [from Morris Motors] was a set of silver candelabras, but the jig and toolmakers made a tobacco cabinet for him. He ran his fingers over the beautiful joinery—and there were tears in his eyes."

Howlett wrote of the incident in his autobiography: "It made me angry, the way [Lord] Nuffield used up people, even people who had been warned . . . as I'd warned [Oliver] Boden. Only Len Lord and Arthur Rowse stood up to him and survived as free agents. Poor little Cecil Kimber was just one of the great majority who seemed to give their very souls into his keeping, and [Lord] Nuffield, for all his great public benefactions, was a rather awkward god in private."

Howlett related how Kimber brought the news to the Manor House at Fyfield, where Howlett, his wife Muriel, and Gladys Hamilton were staying in a top-floor flat.

> *He was white-faced and stunned. He just couldn't understand why he deserved the sack. "You haven't deserved it," I told him, "there must be some mistake; Morris can't do this to you. Why don't you go and have it out with him?" But he came back from the interview more bewildered than ever. "He had me in his office for an hour, John, but he wouldn't let me get anywhere near the issue I'd come to ask him about. He talked generalities for an hour, John. He just kept me talking about things he wasn't interested in to stop me from saying what I'd come to say. And then he indicated it was time for me to go. He evaded me; he tiptoed all around me; he didn't give me a chance.*

Much of the later wartime activity involved clearing paths and creating new roadways suitable for moving heavy equipment around the MG works. Here a new concrete roadway is being constructed by Oxford builder Kingerlees. *Adrian Goodenough*

LEFT: When the war began, factories gave up their heavy trucks and tractors for use in the war effort. The ever-resourceful staff at MG devised this homemade tractor, known affectionately as "Bitsy" (for "bits of this, bits of that"), to make up for the absent support vehicles. Jack Crooke is behind the wheel, while Syd Enever in the dark jacket is standing at center rear. *Enever Family Archive*

BELOW: MG works took on several wartime manufacturing projects, such as building wings for the Hawker Tempest. One of the fastest Allied planes of World War II, the Tempest required some judicious tidying inside the factory just to allow for moving its bulky wings through the production space. *Adrian Goodenough*

Howlett was livid. "But you can't let him get away with that; it's wrongful dismissal—you ought to sue him, make him see reason." But Kimber was not swayed.

> *I don't think I could do a thing like that to him, not to Morris, not after all this time. . . . Nearly twenty years, and me a director for six of them. You see, John, he gave me the chance to be someone in the motor trade, none of us would have been up to much without him. And if he thinks it's right to put me back on the market now, well I must admit he puts me back with better prospects than I had when he took up with me. I ought to be able to find a job easily enough, it's just the way he's gone about doing this to me, that's what's so difficult to take.*

Kimber went to Charlesworth, the coachbuilding firm he knew well as the manufacturers of the open bodies for Morris's SA, VA, and WA cars, before taking a position as works director at Specialloid Pistons in London.

Here his story ends, sadly: He was killed on February 4, 1945, in an accident involving an overloaded train. Kimber was one of two passengers who died. He was 56 years old.

As Jean Cook related her story to the author, she told that she never recovered from the loss. She had just begun to get to know her father, but

war service in the Women's Royal Air Force had meant seeing her father infrequently during the war. "I went to the crematorium at Edgware—I'd never been to such a place before—and I felt that I hadn't been able to say goodbye."

For a final word on Kimber, we turn to John Howlett: "When [Lord] Nuffield turned around at the end of the war, when the motor industry was in a ferment of innovation, and said 'what we need now is a Cecil Kimber,' nobody could provide him with one."

ABINGDON'S WAR STORY

Notwithstanding the shock and upset of Kimber's departure, there was "a war on," as he himself insisted. Life at Abingdon continued: In addition to the Albemarle contract, the factory took in work on designing and refurbishing tanks and other armaments, even rifles. Other aircraft contracts included equipping the legendary Rolls-Royce Merlin engines with mounting frames to allow them to be fitted into aircraft. To aid in moving material about—including from the MG factory at one end of town to the St. Helen's building at the other—Syd Enever and his colleagues in development created a hybrid tractor using a T-Type chassis, christened "Bitsy," which earned an EX number in its own right: EX167.

Abingdon-on-Thames did not suffer greatly from the Luftwaffe's raids—records show that twenty-three high explosives were dropped in the area during the peak period of bombing, between October 1941 and the following June—but a fire at the West St Helen's Street factory in 1944 did calamitous damage. The conflagration destroyed many of the stockpiled parts, including some of the key components from Goldie Gardner's record car (the car itself emerged unscathed) and several of the factory jigs that had been carefully stored there awaiting happier times. Thankfully, those times were just months away.

PLANNING FOR PEACETIME

By 1943, the tide of war was turning. Although factories' priority remained military work, there was enough scope to consider production once the fighting was over. For MG, the departure of Cecil Kimber had left a vacuum. In his

A group of MG factory workers proudly celebrate the assembly of their one-hundredth Albemarle nose cone. *Author archive*

LEFT: One small part of the complex work undertaken in conjunction with assembling the spar assemblies for the Hawker Tempest Mark V's wings. These were in the shape of an ellipse, which made it possible to reach the desired high speed at altitude. In June 1944, these aircraft were crucial to the RAF's part in the invasion of Normandy. *Adrian Goodenough*

BELOW: Work on tanks at Abingdon included conversions, repairs, and rebuilds. *Adrian Goodenough*

place came H. A. (Harold) Ryder, one of Lord Nuffield's longtime associates drawn into the Morris orbit though the assimilation of Osberton Radiators at Bainton Road, the MG assembly plant from 1925 to 1927. Ryder was certainly no evangelist for either sports cars or motorsports, but he was a stout supporter of the MG factory and fought to support it as far as he could.

By this stage, Miles Thomas as vice chairman of Morris Motors was some way into devising the postwar Morris Motors product plan (at least as far as Lord Nuffield would let him). The model range was conceived around the Mosquito saloon car, with Alec Issigonis as the key designer on the project. This car would eventually appear as the postwar Morris Minor. As the smallest model, it was originally designed with a brand-new flat-four engine in mind and would have opened the door for bigger models with inline four- and six-cylinder

engines, covering the Morris and Wolseley portfolios. The plan was somewhat flawed by reports that the flat four suffered from crankshaft problems.

As for MG, however, the future seemed less clear. In Thomas's eyes the classically styled sports cars would be old hat in the postwar period of optimism and recovery, and he felt that the Abingdon factory might be put to better use. Thomas was therefore thinking of adapting the Morris Mosquito saloon further to create an open-topped MG 1100 Roadster or MG Major variant, in addition to the Wolseley Wasp. This was partially due to Lord Nuffield's ambivalence to Alec Issigonis's concept: He had likened the Mosquito prototype to a poached egg, which led to widening the production Minor. Instead, Lord Nuffield favored continuing the upright Morris line of the prewar period.

Harold Ryder was no substitute for Cecil Kimber, but he did fight in Abingdon's corner when he could. In response to a memo from Thomas in

March 1943, Ryder championed the partly developed 1¼-liter MG saloon, which would appear as the Y-Type, as a candidate for Abingdon once hostilities were over. Seven months later, Ryder wrote to Thomas: "Because MG is a small factory, it makes sense to put Nuffield Mechanizations at Cowley, and reinforce the policy of all MG design to happen at Cowley under Mr. Oak and the Drawing Office and Experimental Department at Cowley. If MG is not allowed by government to make cars after the war, the factory can be used to make parts for other cars."

On February 2, 1944, Ryder again wrote to Thomas, reporting on a meeting with what he called rather benevolently the "Executives" at MG:

> *I must say, they made a very good case for the introduction of a Midget. I think you will appreciate that this company was built up on a Midget car, and I feel that provided we run a "bread and butter" line, which is an improvement on the standard, we could economically produce a Midget car: at least for a few years after the war in order not to give the public the impression we were falling out of the market which had given us such a good name. I have therefore contacted Mr. Oak, and he informs me that the chassis of the 1¼-litre can be readily used for a Midget, and he is investigating the possibility of body styles and types, using as much as possible, the standard panels.*

As we shall see in the following chapter, what the market wanted and could support in the postwar export-focused period would drive the direction of the MG sports car, and the idea quoted by Ryder was at the heart of it.

LEFT: Harold Ryder was an old associate of Lord Nuffield, having worked at the business that became Morris Radiators Branch. In the wake of Kimber's departure, he assumed overall responsibility for MG. *Author archive*

BELOW: One of MG's special projects: converting Centaur tanks into armor-plated bulldozers that could clear rubble from streets in war-ravaged Europe. They were designed to shield the driver from enemy sniper fire. MG's Syd Enever spearheaded the Centaur Dozer conversion. *Adrian Goodenough*

OPPOSITE TOP: The MG TB (CJB 59) again, along with a prototype MG Ten, which would eventually appear as the postwar MG Y-Type 1.25-liter saloon. *Author archive*

OPPOSITE BOTTOM: Tank assembly at Abingdon. *Author archive*

MG
CJB 59

4

Postwar Recovery

PICKING UP THE PIECES: MG TC AND Y-TYPE

The Allies were thinking ahead to the end of hostilities in Europe by 1943, though the global conflict would grind on for at least two years—and its aftermath lasted well into the following decade. Even so, by 1945, Great Britain's carmakers were chomping at the bit to restart production. The military contracts that had sustained them through wartime would soon evaporate, and the absence of new cars during wartime meant the market would be hungry for product.

It would take time for full production to resume, though, as MG employee Jimmy Cox recalled. MG's continued manufacture of Neptune amphibious tanks was one of many instances of bureaucratic nonsense that tied up the factory's resources. Once built, the tanks were inspected, tested, and then scrapped because the contract had not completed its course.

At the war's conclusion, the first effort undertaken by MG was the renewal of the Midget line, last seen as the TB for a brief period in 1939. The postwar version was the new TC Midget, which, largely because the Luftwaffe bombing had destroyed the original tooling in Coventry, now required a new chassis and jigs and allowed for changes based on lessons already learned in service, such as suspension weaknesses. Work on the TC began in earnest in spring 1945, with the drawing work largely undertaken at Cowley by Jim Stimson, a draughtsman who had worked on the Mosquito. Stimson later moved across to Abingdon to work with Syd Enever.

OPPOSITE: Nuffield Exports' European representative (and former MG publicity manager) George Tuck is pictured at the wheel of this MG TB Midget somewhere near Abingdon in 1946. This car evolved into the TC prototype ("TC0251"). Tuck originally coined the famous MG slogan "Safety Fast!" *Author archive*

RIGHT: A 1946 MG TC Midget alongside the kind of petrol pump that its first owner might have seen when they filled up. The color scheme is Sequoia Cream with red trim and matching radiator grille vans. *James Mann*

BELOW: MG's exports to the United States picked up where they had left off before the war, with the Collier Brothers' Motor Sports Inc. of New York. These five TC Midgets are about to be loaded into the Number 2 Hold of the liner *Queen Elizabeth* at Southampton in 1946. *Author archive*

Overall, changes to the Midget were modest but useful: a slightly wider cockpit, improved front suspension, and a single twelve-volt battery in place of the previous twin six-volts. The first preproduction TC was car number 0251 (named for "Abingdon 251," the factory telephone switchboard number). It was completed in September 1945, and before long the first batch of cars was rolling off the production line.

Home-market sales were severely depressed—even before the heat of the coming national export drive—and many of the new cars were sold abroad, mostly to traditional markets in the British Empire, as there was no provision for left-hand-drive vehicles. Only a handful of TC Midgets went to the United States, where demand for the cars was insignificant, while initial sales of the TC to markets such as Australia outstripped most other exports.

The war had been costly for all nations involved in the conflict. In Great Britain's case, while victory of the Allies against the Axis was obviously the desired outcome, the bills attached to this successful effort were now coming due. Apart from the need to rebuild damaged cities and factories, there was the enormous debt tied to President Roosevelt's Lend-Lease scheme, which had provided vitally needed help with war materiel as well as more basic needs. The amount granted to the United Kingdom through Lend-Lease was about $31.4 billion (roughly $750 billion in 2023) and, unsurprisingly, much of the "loaned" equipment was not available for return.

Despite having brought his nation through World War II, Winston Churchill and his Conservative party were rejected by voters in the first postwar election in 1946. Under the new Labour prime minister, Clement Attlee, the political rapport between the U.K. and the United States began to cool. There was growing awareness of the need to export goods and services to earn foreign currency, especially the U.S. dollar.

In August 1946, the new Ministry of Information produced an animated film, *Export or Die*, whose message to the public was to restrain their desire for luxury items denied them for the past six years and allow those goods to be exported to benefit the national economy. Exports, the narrator explained, generated money to buy food and raw materials such as timber. The implication was that it was far better, surely, to ship shiny new MGs and other British cars abroad while those back home made do with a car laid up during the war years. "We must sell the things we like to buy the things we need," the narrator implored.

Around the same time, Nuffield Exports launched a new magazine aimed at internal colleagues as well as distributors and agents. Early issues focused on exports to the traditional Empire and Colonial markets, but it was clear there was more to be done. Meanwhile, the current U.K. government, far more interventionist than its prewar predecessor, encouraged British carmakers to rationalize their car ranges, implying that this would give exports a better chance of success. The government also dropped dark hints that material supplies might be rationed. U.K. industrialists looked on in concern as the Labour government moved to nationalize some major industries and take them into public ownership. Surely such a plan would never be applied to the car industry, they hoped.

Matters were hardly helped by bitingly cold and severe winter weather in the U.K. in 1946–1947, which precipitated terrible fuel, power, and material shortages. These issues proved damaging to the car industry along with most aspects of life. Stresses and strains of this kind were viewed as threats by Britain's major carmakers, and in this turbulent time many projects were either reshaped or abandoned as investment was refocused.

MG's modest U.S. sales in the postwar period were largely handled by Motor Sport Inc. on New York's Fifth Avenue, a business founded by Miles and Sam Collier, who had imported and raced MGs before the war. As a publicity stunt, they arranged with Nuffield Exports to ship five brand-new TC Midgets from Southampton to New York on the garage deck of the *RMS Queen Elizabeth*, at the time the world's largest passenger liner. Having been repurposed in 1940 as a troop carrier, the *Queen Elizabeth* had undergone a costly postwar refit and embarked on its maiden voyage as a commercial passenger liner on October 16, 1946. The TCs crossed the Atlantic on the *Queen Elizabeth*'s five-and-half-day voyage in the early days of 1947, a move proclaimed as "further proof of the determination of Nuffield Exports to ship their vehicles overseas in the fastest and most efficient manner possible." It was a bold gesture, but it was no more than a drop in the ocean. A better answer to sales expansion was on the horizon.

With the entire U.K. automotive industry struggling, the climate was right for change. In November 1947, Miles Thomas left the

The 1.25-liter Y-Type saloon. Despite being hailed as a fresh product, the car nevertheless harked back to prewar styling practices, a consequence in part of the bodywork being derived from that of the old Morris M Series. *James Mann*

TOP LEFT: The tall figure seen alongside the MG 1.25-liter is Cecil Cousins, MG's works manager. These are the first two production models; in the foreground is the first to be shipped to Australia, and behind it a car destined for Malaya (present-day Malaysia). Of the three men behind, the one in the center is Hector G. Cox, sales manager of the MG Car Company at the time. *Author archive*

TOP RIGHT: Jocelyn Hambro was a member of the British Merchant Bank. Fired with enthusiasm for securing export business in the United States, he secured a lucrative arrangement with the Nuffield Organization, which included the distribution of MG sports cars. *Author archive*

ABOVE: This 1947 photograph appeared in the January 1948 *Nuffield News Exchange* as part of an article celebrating the company's export drive. Here the drivers make their first stop on the way from the Cowley depot, along various roads, to London's Royal Albert Dock. The hotel, now a pub, bar, and restaurant, looks mostly the same in 2023. *Author archive*

Nuffield Organization—another victim of Lord Nuffield's frequent vicissitudes in his relationships with his own management. Thomas had become involved in the Labour government's Colonial Development Corporation. His boss saw this as diverting attention from his Nuffield responsibilities. In Thomas's wake, some of the remaining plans for oddball MG and Wolseley offshoots of the Morris Minor were soon abandoned.

A month later, R. F. "Reggie" Hanks replaced Thomas as the new deputy director. Hanks got rid of more of the old guard, many of whom were past retirement age, including Harold Ryder. Followed by others, such as sales director Donald Harrison, Hanks swiftly embarked on a series of fact-finding visits to overseas markets, in particular to the United States.

Hanks hoped that British small saloon cars would do well in a U.S. market where the domestic car industry was still recovering and struggling to meet demand. This proved to be wishful thinking, as the few cars Hanks could offer were unsuited to local conditions and, worse, they were expensive. Hanks described the situation after one of his trips:

> *It does seem so silly that, in travelling all these miles, the only thing I am able to talk about in any seriousness, is the TC Midget which was virtually obsolete many, many years ago, and yet there are a certain class of people who like it! Put the same type of body on the 1¼-litre chassis and we have got something.*

But price must be reduced. . . . If we really want to come into the market, let us offer something the American Factories do not offer to the public (and, I suggest, never can).

MG's first properly new postwar car appeared in May 1947: the MG 1¼-Litre or Y-Type four-door saloon. This was in effect the MG Ten, originally intended for the 1940 Motor Show, had there been one. It could be said to be the last MG to have benefited from Cecil Kimber's influence, even though the Morris Eight Series E's origins were hard to disguise. The car's styling had obvious prewar roots, with its separate headlamps; narrow, tapered body; and traditional MG radiator grille. Though unlikely to set the world on fire, the Y-Type would arguably meet its intended clientele's slightly traditionalist tastes.

The Y-Type was different in many ways from the TC. For the Midgets, the body shells came partially finished, but the bodies of the new saloons arrived at the Abingdon factory almost totally bare. Luckily, the factory had already taken on a trimmer in 1942, when Jack Hayward had arrived to teach welding to the women working on the production line. Hayward stayed on after the war and, when Cecil Cousins found out that he was a trimmer, asked Hayward to help with the Y-Type.

The new saloon was highly anticipated. For the English and Welsh dealer launch, the service bay at Abingdon was bedecked patriotically with red-white-and-blue flowers and the Union Jack above a line of the new Y-Types. With this blatantly U.K.-centric decoration, there was perhaps little hope that it would be met with enthusiasm in the United States.

JOCELYN HAMBRO ARRIVES

By the latter half of the 1940s, it was evident that Austin was ahead in the export game. After starting his new tenure at the Longbridge factory, Len Lord had wasted little time in formulating Austin's plans for North American exports. These plans, hobbled only temporarily by the war, included the launch of the new Austin A40 Devon in New York, which took place in September 1948. It was a clear demonstration of Austin's ability to take advantage of having continued some car and related manufacture through the war period. The switch back to car manufacture cost Morris time, and the company remained on the back foot as its rival sped ahead.

North American visits by Miles Thomas, Reginald Hanks, and others reinforced the depressing reality of how far behind Nuffield Exports' efforts were in chasing U.S. dollars. Events like sending TC Midgets over on the *Queen Elizabeth* were fine for publicity, but before long Austin could proudly boast of figures like their one-hundred-thousandth export car (June 1948) and postwar U.S. sales of more than $22 million (December 1948). Austin also laid plans to build cars at a new factory in Hamilton, Ontario; Nuffield had nothing to match this. Perhaps these issues, as well as other rivalries, prompted the announcement in October 1948 of a short-lived collaboration between Austin and Nuffield.

Aid came in the form of a dynamic merchant banker and horse breeder, not yet thirty years old, named Jocelyn Olaf Hambro. Recognizing that the U.K. government's "Export or Die" message could be turned to profit, and armed with a $10,000 line of credit from the Bank of England in addition to the capital of his family's bank, he set up the Hambro Trading Corporation of America in New Orleans in June 1946. On the face of it, New Orleans seemed an unusual choice of headquarters, but Hambro was keen to reach markets in the American Midwest rather than tread water in prosperous but saturated New York.

Led by sales manager Owen Slater, Hambro's small team was inventive in terms of the "theatricals" they pursued. Here we see the TC Midget being shown to Texan royalty at the summer 1947 State Fair. The car also joined a parade featuring the 1947 Cotton Queen. *Author archive*

Hambro bought up various assets—early business interests included English pottery and James motorcycles—and then sold them on credit terms to local businesses, none of whom had heard of Hambro but seemed receptive to his persuasive powers. Motorcycle distribution took off, with more than fifty thousand units sold, although once the domestic U.S. car industry recovered, customers for motorcycles turned to other transportation products. Hambro took the entrepreneurial gamble of buying six TC Midgets, which he promptly sold in the Midwest. As Hambro explained to his biographer, Andrew St. George, "I used to ask [Hambro Bank's] Owen Slater, who was in charge of finances, if we were all right for cash, and he used to do a few sums on a piece of paper, and say 'yes, we're all right.' I was never allowed to see the figures!"

Giving the Texas State Fair's Cotton Queen a ride in a TC through the center of Dallas, followed by a demonstration to Texas governor Beauford Jester at the fair in the summer of 1947, was quite a publicity coup, but Hambro was probably less thrilled when the mayor of Laredo, Hugh S. Cluck, characterized the MG as "first cousin of a malted milk machine." Still, publicizing the MG sports car beyond the usual locus of New York was a positive step, which Hambro and advertising associate Patrick Dolan further leveraged when they launched an MG Car Club in California, garnering coverage in *Life* magazine.

Prior to the ascendancy of the U.S. market, MGs were exported in larger numbers to Switzerland, Australia, and Belgium. "Bound for Switzerland" was the title for this photo, which depicts a line of MG TC Midgets and a single three-seater Riley Roadster on the jetty at Dover, about to be driven aboard the Hampton ferry boat. *Author archive*

The Dallas event was covered in the December 1947 issue of Nuffield Exports' *News Exchange* magazine. Reviewing the article now, it is interesting to note that the TC had been fitted with a substantial front bumper, although the car remained resolutely right-hand-drive. The U.S. export version of the TC became the EX-U model, with a few concessions to American demands that included twin wind-tone horns, modified lighting, and front and rear bumpers. In its next issue, for January 1948, the *News Exchange* reported the appointment of Leslie Marchant as American sales representative for MG and Harry Rummins as his service counterpart. Both men came from the MG works at Abingdon. This news came on the heels of cables from Nuffield Exports general sales manager Donald Harrison regarding his coast-to-coast trade tour of the United States.

> *Mr. Harrison's cables showed that there is a definite market for certain products of the Nuffield range throughout the USA, particularly the open MG Midget sports car, for which large orders have already been received. It will be the task of Marchant and Rummins, in close co-operation with Mr. Owen Slater of the Hambro Trading Co. Inc. They reached New Orleans a few days ahead of the first big consignment of fifty MGs which left London Docks recently.*

A significant challenge for the new organization was funding and vetting suitable distributors. Unsurprisingly, domestic car dealers were seldom interested, so Hambro made deals with a panoply of oddballs including a business in Dallas selling Seberg jukeboxes, the Schlitz Beer distributor in Chicago, and the Philco Radio distributor in Southern California. It was exciting in a sort of pioneer-spirit way, though as Hambro told his biographer, "Our

Welsh-born Ray Milland was a famous Hollywood movie star at the time. Here he takes personal delivery of a new MG TC Midget at the Nuffield Exports offices at Cowley in 1948. This was one of many such publicity exercises where celebrities of the day were pleased to be seen with the "must have" MG sports car. *Author archive*

car distributors, based on beer, radios, and bootlegging, didn't really understand how to treat my wife, and she was constantly pinched on the bottom by all of them." Nevertheless, Hambro and Nuffield earned some £61,000 in the first year and built the business steadily. It seemed at times an eccentric arrangement, but Austin was trying to manage its own sales as a wholly owned operation with less success. Hambro Bank did especially well thanks to low overhead, with an operation consisting of only a New York office and six employees.

News Exchange further reported that "a very definite market has been opened up for the MG Series TC Midgets—one of Britain's most successful sporting cars. Already these, in large quantities, together with some Morris Ten Series M saloons, have left this country for New York, Los Angeles and New Orleans. . . . Added to this, very large shipments have been going to Latin America, Canada, Bermuda and Barbados."

In February 1948, U.K. government minister for supply George Strauss announced that steel would be reserved for supply to those

Hambro's Other Ventures

Not everything that Jocelyn Hambro tried succeeded as well as the James motorcycles and the Nuffield (and later BMC) ventures. The Hambro House of Design sold contemporary British and Scandinavian domestic goods such as glassware and linen. That business lost money and was closed in 1956. Even less successful was an ill-starred attempt to sell Scottish kippers (the Queen of Scots) to New Yorkers. The smell of rotting fish provided a lingering reminder of that failure. Another venture involved a warehouse full of honey; the product fermented and exploded due to heat, covering walls, floor, and ceiling in a sticky mess. Today, Hambro would be lauded as a serial entrepreneur.

TOP: A contemporary of the new Morris Minor was the open-topped MG four-seat tourer version of the 1.25-liter MG Y-Type, known formally as the YT. The Nuffield Organization stressed that the car had been "designed for the export market." *Author archive*

ABOVE: As far as MG's parent body, the Nuffield Organization, was concerned, the most important new products for the end of the 1940s were the all-new Morris Minor, Oxford, and Six models. The Minor was certainly one of the stars of the 1948 Earls Court Motor Show, the first postwar London event of its kind. Initially launched with the low headlamps seen in this photo, this configuration ran afoul of California safety standards. *Author archive*

car manufacturers exporting more than 75 percent of their output. The problem for Nuffield was that, other than MG, little of its automotive output was suitable for the new target markets overseas. This did not prevent bold efforts, however. For example, from the *News Exchange* of August 1948 we learn of a new export record for Nuffield Exports that had been set in June, "when 1,553 cars and trucks, valued at £500,000, were shipped in one working week of five days. . . . among the shipments to hard currency countries were 100 MGs which left Manchester docks on one boat for Los Angeles, and a similar quantity of cars and trucks dispatched in another vessel to Canada from London docks."

In U.S. dollar market export terms, MG was the Nuffield Organization's one obvious success story. In autumn 1948, Nuffield sent Wolseley's general sales manager, C. H. Fison, to the United States "for some months" to see if he could drum up Wolseley sales. He could not.

The debut appearance of the Issigonis Morris Minor at the first postwar Earls Court Motor Show of October 1948 offered hope. The Minor was part of a now largely forgotten line that also covered larger Morris Oxford and Six lines, as well as various Wolseleys—but no MGs. Also at the show was the MG YT Tourer, another early Nuffield postwar effort to chase after export currency.

One key problem was that MG's fate was now largely steered from Cowley. This was mostly because, to many of the people at that mighty works, Morrises and Wolseleys were considered far more relevant and important. Over 1948 and 1949, MG's export potential would be considered, but much of the thinking was muddled and parochial—even if the sales side could rightly caution against killing the proverbial golden goose for which Abingdon was reaping glory.

NUFFIELD PONDERS MG'S FUTURE

The postwar appeal of the MG sports car's old-fashioned style bemused Nuffield Organization management, who were looking toward the future of automotive design trends. On the other hand, Lord Nuffield himself was no fan of what he saw as pernicious modernity (recall that he had disliked the original shape that spawned the new Morris Minor).

In his later years he preferred his trusty 1939 Wolseley Eight, whose body closely resembled that of the new MG Y-Type. It seems likely that, in his mind, America's embrace of the MG Midget's traditional style—and steady sales of the Y-Type—were vindication of his opinions. All that said, Lord Nuffield's influence on new model styling was waning, even if from time to time he could still flex his opinions.

Managers who had steered the company through the decade were leaving—or being fired. Some of this may have been due to the company's lack of support for innovative designs: We have seen how Miles Thomas cooked up MG versions of the Minor family, including the MG Midget Major (DO 926), though his ideas went nowhere. After Thomas and Ryder departed in late 1947, Hanks and

LEFT: Also involved in directing MG affairs was Sidney Victor (S. V.) Smith. Reporting to Reggie Hanks, he kept a close eye on Abingdon's affairs and was known (behind his back) as "Hitler Smith" on account of his brushlike mustache. *Author archive*

BELOW: Somewhere in this showroom lineup at Morris Motors, Cowley, lurks a prototype for a new MG sports roadster, the Midget Major (DO926), which would have been derived from Morris Minor components. Despite ambitions from the Cowley side, the concept went nowhere. It was not the last such design scoped with little Abingdon input. *Author archive*

The MG TD was an evolution in terms of Midget styling, but it was more modern than the outgoing TC Midget in terms of chassis and suspension. To the horror of many contemporary MG sports car enthusiasts, it had also dispensed with the TC's spindly 19-inch (48 cm) center-lock wire wheels. For the first time, however, export customers could buy a left-hand-drive MG sports car. *Author archive*

new divisional director Sydney (S. V.) Smith took charge, while technical direction with oversight of MG fell to A. V. Oak. Plant management at Abingdon remained with George Propert, though he would retire in July 1949.

Hanks had been a premium apprentice at the Great Western Railway works at Swindon, served in the Royal Army Service Corps in World War I, and later joined Morris Motors Service in 1922. He rose through the ranks, becoming chief inspector of the Cars Branch in 1936 and production manager in 1938. At the start of World War II, he became manager of the RAF Civilian Repair Organisation in 1940, overseeing repairs of Spitfires and Hurricane fighter planes at Cowley, then from 1941 onward held the position of general manager of Nuffield Mechanizations, where tanks were built. Finally, at the close of the war, Hanks was appointed general manager of Nuffield Exports in July 1945. Hanks was in a good position to slip into the role vacated by Miles Thomas.

S. V. Smith was a former Wolseley apprentice who had also served in World War I. He remained with Wolseley when Morris acquired the business in 1927 and subsequently moved to Cowley in 1933, becoming works manager three years later. He was appointed to the Morris Motors Board in late 1947 before Thomas's departure and the subsequent purge of older directors. For Hanks and Smith, the focus was on Cowley and Coventry, but they also had to consider the sprawling outliers of the Nuffield Organization such as MG, Riley, Wolseley, and, to a lesser extent, Morris Commercial Cars.

MG and Riley both relied on the traditional coachbuilding skills of Morris Motors Bodies Branch, which, like Riley, was based in Coventry. In short order, Hanks and his board colleagues formulated the idea of bringing MG and Riley together into one unit. Their first thought was to bring everything under one roof in Coventry and turn the Abingdon plant over as a satellite to serve nearby Cowley. In this same period, there were various exercises to develop possible new MG cars—a roadster, a coupe, and a two-door saloon—using Riley engines and more modern body styling, but mirroring the coach-built construction of the RM Rileys. These exercises, under Cowley codes DO 963, 965, and 967, were developed from June to October 1948. Clearly ideas of rationalization were behind the exercise, but once again there seemed to be little evidence that management understood the demands of the growing U.S. export market.

In December 1948, the Morris Motors Board considered a confidential "Nuffield factories re-organisation scheme," which included a number of key moves. Wolseley car making, at Ward End, Birmingham, would move to Cowley, and Morris Commercial Cars would assume the space vacated at Ward End. Engine manufacture, in the form of the Morris Motors Engines Branch, would be chiefly focused on Courthouse Green in Coventry. Lastly, the old Radiators Branch would decamp from Bainton Road to Birmingham and Llanelli. The big change for MG came under what was called "Phase 5," which set out that "MG Car Company and Rileys will combine production resources at the Riley factory in Coventry. This will create a tidy bloc comprising our two 'specialist' cars with Bodies Branch on their door-step. All three factories will be under common Management."

"Phase 6" involved the sales and marketing function: "The Sales Departments of Morris, Wolseley, MG, Riley and Marine and Industrial Engines will be brought together at Cowley under the direct control of the Sales Director." This position was held by Donald Harrison, who had been one of the visitors to the United States on fact-finding missions. "Phase 6A" related to servicing, with aspects of MG and Riley operations being merged at the Riley Works in Coventry.

These plans were supposedly confidential, though some detail inevitably leaked out. On the principle that bad news travels fast, many of the leading lights at MG soon got wind of what was being proposed. George Propert, Cecil Cousins, and John Thornley led the MG team to mount a clever but careful campaign to influence the man tasked with planning the chess moves, Morris Motors' chief planning engineer Tom Richardson. The hope was that MG would remain at Abingdon, and, if anything, Riley, with its even more archaic car-making techniques, should be brought down to the MG works.

The MG men proved successful in their task. In January 1949, the Morris Motors Board received a fresh plan that set out, among other things, that "Riley [will] move to the MG premises at Abingdon where assembly of the two marques will be combined." The chief reasons given for this reversal of the original

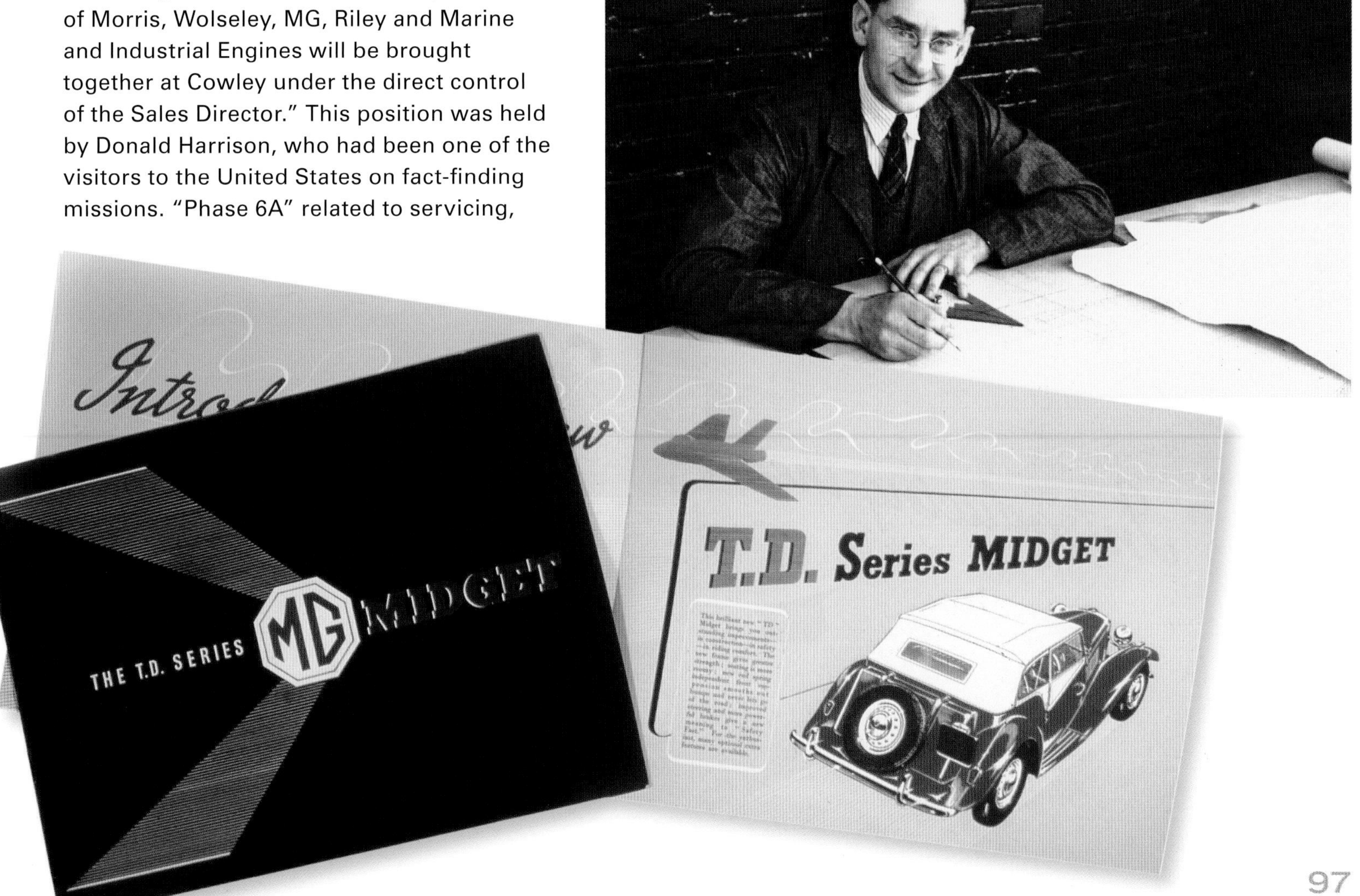

BELOW: Jim O'Neill was one of the team at Morris Motors Cowley responsible for development of the TD Midget. Syd Enever later recruited O'Neill to form part of his MG design team at Abingdon. *Denis Williams*

BOTTOM: Contemporary sales literature for the MG TD Midget. *James Mann*

Some U.S. enthusiasts were clamoring for a more modern MG style, so a handful of enterprising businesses sought to answer that call. This elegant solution was developed by S. H. "Wacky" Arnolt, who commissioned this elegant body style from Bertone. The Arnolt MG TD was also built in coupe guise. *James Mann*

plan were a lack of space at the Riley plant and the need for some accommodation by the Morris Motors Engines Branch. There was a comment that "future deliberation [would] be given to the prospect of transferring Bodies Branch to the Cowley region"—an idea that never came to pass—and MG and Riley Service would be combined at Abingdon.

Whatever changes were to be implemented, George Propert would not be there to see them go into effect: He retired on July 20, 1949. His place was taken by Riley man Jack Tatlow, and before long there were Rileys coming off the lines at Abingdon alongside the MGs. One of Tatlow's early MG-related tasks was to see the first new postwar MG sports car into production.

Around the same time, a new chief designer appeared at Cowley. Gerald Palmer was tasked with overseeing the creation of a new range of MG and Riley (and, before long, Wolseley) saloons. Palmer would also preside over two key models, the new MG Magnette and the Riley Pathfinder, as well as a distinctive new engine that would only ever appear in MG guise. But that is a story for later in this chapter.

DESIGNED FOR THE UNITED STATES: MG TD MIDGET

The first step toward a new postwar MG sports car was a good one. The TD Midget took many of the TC's mechanical attributes (in particular its strong and tunable XPAG engine) and its classic styling with distinctive double-humped front tonneau panel, upright radiator grille, folding windscreen, "suicide doors," and compact size. To this were added more modern suspension, smaller-diameter wheels, and, for the first time in an export MG Midget, the availability of left-hand-drive steering. Much of the design work was undertaken by a young engineering draughtsman at Cowley named Jim O'Neill, who in due course would take his place as one of the first members of the design team established under Syd Enever at Abingdon.

We have seen already in chapter 3 how the basic concept for what would become the TD came from Cecil Cousins and his colleagues at Abingdon, who had "cut and shut" a Y-Type chassis to create the basic concept. This was only a static demonstration of a promising idea, though, as much more work was necessary to create something that would work in the real world. O'Neill later told the author that the brief he received from Morris Motors' assistant chief body draughtsman Tom Ramsay. A lead on the project who had been responsible for both the TC and Y-Type body shells, Ramsay was to produce a squatter, less angular body shape than the TC, while using the Y-Type's more up-to-date front suspension, which had been designed before the war by Issigonis and detailed by Jack Daniels. The chassis design for the new car was also led by Daniels.

The new MG sports car project was given a Cowley Drawing Office code, DO 968, and thus fell under the purview of Cowley's technical director, Vic Oak. O'Neill found the senior management largely disinterested at first: "Oak used to go around all the Morris drawing boards, but he never even stopped to look at my drawing. I concluded that Morris Motors were just not interested in MG." More encouraging was a visit from John Thornley, who came over from Abingdon to see what Cowley was cooking up for them to build. Thornley was still service manager, though he would soon be promoted.

O'Neill told the author that he did not know how much Thornley understood when he reviewed the complicated body layout drawings, "but he made a point of asking if we could lower the tonneau line by half an inch [1.25 cm]. This would reduce the height above the rear wing and give an even more squat appearance. I did this while he was still there—and sure enough, it improved the shape considerably. With a 'leave to go, Sir,' he departed." The design was taken to Morris Motors Bodies Branch at Coventry and translated into a prototype body, complete with traditional ash frame and steel panels. More problematic, however, was the chassis.

Mounted on the bare chassis, the prototype of what would become the TD looked good, but the first road test soon revealed some fundamental issues. The road test report stated: "handling good but impossible to read instruments at any speed." It was the classic

ABOVE: Jack Duckham (*right*) celebrates with Goldie Gardner on Gardner's achieving an International Class F record of 137.4 miles per hour (221.12 kilometers per hour) in Utah in 1951. Duckham Oil was a key sponsor of Gardner's record attempts. *Author archive*

BELOW: In 1946, Goldie Gardner took up where he'd left off in 1939. Here he is in Monte Carlo, on his way to Italy, flanked by his trusty Abingdon lieutenants, Reg Jackson and Syd Enever. *Enever Family Archive*

By 1951, Gardner had decided that the usual European roads and race circuits were no longer sufficient for the higher speeds he was chasing. The record breaking therefore shifted toward the famous salt flats at Bonneville, Utah. Here Goldie talks to another record breaker, Ab Jenkins, known as the "Mormon Meteor." *Enever Family Archive*

ABOVE: The finished, special-bodied MG TD, EX172, was road-registered as UMG400. *Enever Family Archive*

LEFT: Gerald Palmer, MG and Riley chief designer from 1949 to 1955. He was forced to resign due to a clash with Leonard Lord. As well as the DO1049 (and related DO1048) sports car concepts, he presided over the MG Z-Type Magnette and the larger Riley Pathfinder (and Wolseley 6/90), and he conceived the twin-cam version of the new BMC B-Series engine. *Author Archive*

BELOW: Syd Enever's second effort to follow the T-Type MG sports car was unquestionably a superior attempt. It used a brand-new chassis with outboard members, which allowed the driver to sit lower in the car. Styling was sleek and more modern than any previous production MG sports car. Perhaps it drew some inspiration from other new postwar sports cars like the Jaguar XK120, although in truth the shape of the EX175 was clearly an evolution of the earlier EX172. *Author Archive*

problem of scuttle shake, a shuddering vibration due to lessened structural rigidity, which was the bane of many open sports cars of the period.

Thankfully, O'Neill's colleague Tom Ramsay, one of the few Morris Motors people to take much interest in MG affairs, came up with a solution. A sheet of ⅛-inch (20-cm)-thick vertical steel plate from the chassis up to the body dash was welded in place to give the needed extra stability, then a 12-inch (30.5-cm)-square hole was cut in the plate to allow the driver's feet to reach the pedals. This cured the scuttle shake, but it was clumsy. A better fix was found in the form of a tubular "goalpost" welded to the production TD Midget chassis frame in line with the dashboard: a far more practical alternative. This solution was adopted.

As part of the shift of plant functions described above, in 1949 Charles Griffin transferred from Wolseley to become deputy chief experimental engineer for the Nuffield Organization. He told the author that, in his view, the scuttle shake was incompletely cured by the goalpost, which he dubbed the "towel rail." Despite this, MG's marketing men made a virtue of the rollover strength it provided to the car's overall structure.

Meanwhile, the U.K.'s politicians continued the drive for more product export. Partly to stimulate that goal, the government devalued the pound sterling from $4.03 to

ABOVE: When Leonard Lord initially rejected the EX175 for development, Abingdon was forced to undertake a comparatively lightweight face-lift of the three-year-old TD, creating the TF Midget. This vehicle had sweeping wings and a swept-back radiator grille (now decorative rather than functional). The car in the photograph is the prototype of EX177/DO1047. *Author archive*

BELOW: The MG *TF* Midget of 1953 was clearly an evolution of the earlier Midget family rather than the all-new car that Enever would have preferred. Although some of the more progressive sports car enthusiasts may have turned up their noses at the design, it was warmly received by many who regarded it as one of the most attractive T-Type Midgets. *Author archive*

ABOVE: The 1953 Earls Court Motor Show saw the debut of not only the TF Midget, but also the far more contemporary MG Z-Type Magnette (DO1010 was the Cowley project code; this photo shows a prototype in the studio at Cowley). Largely the work of Gerald Palmer, its sleek modern styling was described as "airsmoothed" in the publicity campaign. This quality won the marque of many new friends, even if a vocal minority of traditionalists decried the use of the sacred Magnette name on such a car. *James Mann*

BELOW: Filling the void left by Goldie Gardner's retirement from MG record breaking, Captain George Eyston returned to take his place. By now he was an influential figure at Wakefield Oil, the parent company of Castrol. He is seen here seated in the cockpit of the new EX179 record breaker designed by Enever. Sensibly, Eyston agreed to delegate the actual driving to younger men. *Author archive*

$2.80 on September 18, 1949. MG was in the process of readying their prime candidate to take advantage of the currency change. The first production TD Midget went down the line on November 10 of that year, followed not long after by the ten-thousandth and final TC, chassis number 10,251. It was an auspicious beginning for the TD as it launched on January 18, 1950, exactly four months to the day after the dramatic currency devaluation. Well more than 90 percent of the 4,767 TD Midgets built in 1950 would be exported, with more than half going to the United States.

MG RACING AND MORE RECORD BREAKING

As we saw in chapter 2, Goldie Gardner and his troupe had hoped to continue their European record-breaking escapades had World War II not intervened. No sooner had hostilities ceased than Gardner laid plans to pick up the racing threads. Unfortunately, record runs at Dessau were now out of the question, that space having been swallowed by East Germany. Attention turned to the Brescia-Bergamo road in Italy, with the enthusiastic support of Count Lurani. The trip in July 1946 found the site less than ideal, so Gardner switched his focus to Jabekke in Belgium. There, in October 1946, he achieved 164.722 miles per hour (265.095 kilometers per hour) on the return run of the Flying Kilometer—from 750 cc! This was not matched in a related attempt at the 500 cc record, which failed to beat prewar speeds.

In spring 1947, the Gardner car was shown at the Geneva Salon, and that summer the party returned to Jabekke. There they tallied the Flying Kilometer at 118.043 miles per hour (189.97 kilometers per hour), Flying Mile at 117.105 miles per hour (188.46 kilometers per hour), five kilometers at 114.105 (70.9 miles per hour), and five miles at 110.531 miles per hour (177.88 kilometers per hour). Notwithstanding the Motor Show exhibit in Switzerland, support for the essentially prewar Gardner record car within the Nuffield Organization was beginning to wane. Factory assistance from Abingdon was constrained, and when Gardner asked Nuffield if they had a two-liter engine he could use in 1948, he was told there was nothing suitable. Gardner went to Jaguar instead to secure a prototype four-cylinder version of their new XK twin-cam engine.

LEFT: A key part of the battle to show that the new MG was a genuine sports car, not just a pretty showroom decoration, was the process of getting it to "earn its spurs" in serious motorsports. Here race driver Ken Wharton tries out one of the EX182 prototypes at the famous British Silverstone circuit in the spring of 1955. The new model was described as a "prototype of a possible future production MG sports car," hence the need for subterfuge by referring only to the EX182 project code. *Enever Family Archive*

BELOW: The key team at Abingdon responsible for the EX179 stand beside it ahead of the 1956 record-breaking trip. *Enever Family Archive*

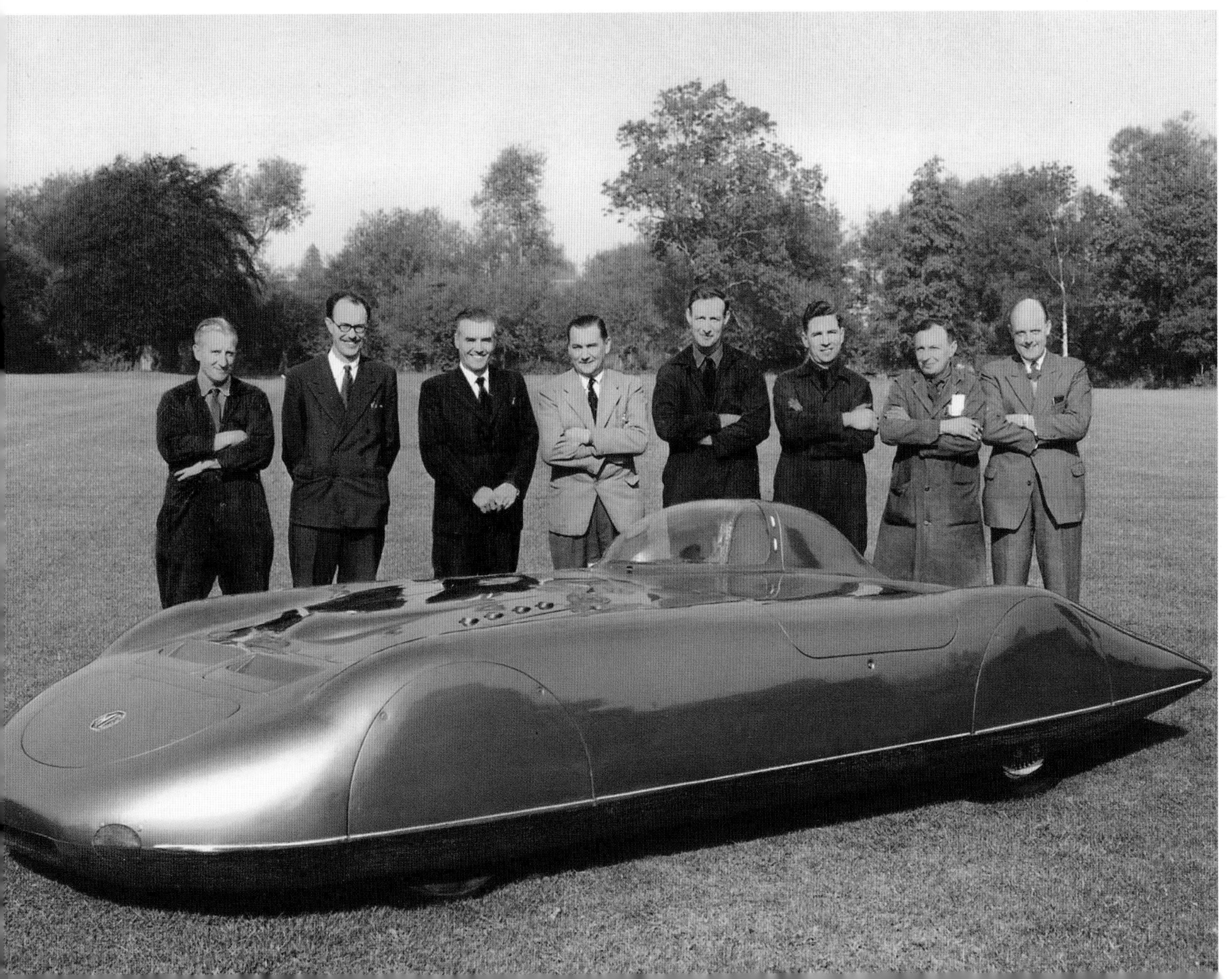

RIGHT: Understandably, few people outside of BMC knew anything about the EX175. That made the task of developing it into a production version somewhat easier, although production delays with the body tooling meant that the debut of the roadgoing car was delayed, allowing the lightweight EX182 cars to be revealed in June 1955 as racing prototypes. This series of side elevations shows how EX175 evolved into EX182 and ultimately the MG Series MGA, the latter with Cowley project code DO1062. ***Author archive***

BELOW: Richard "Dickie" Green and Jimmy Cox stand proudly in front of the MG pit at Le Mans in June 1955. ***Cliff Bray***

The smartly turned-out MG EX182 cars at Le Mans scrutineering prior to the actual race. *Cliff Bray*

The MG name was temporarily out of record-run contention through 1948, but determined voices intervened, and Gardner was soon back in the fold with trips to Jabekke in 1949 and 1950 with greater boldness and vaunting ambition. In 1951, he ventured across the Atlantic to Utah's Bonneville Salt Flats. For the 1951 trip, the car was shipped in the hold of the *Queen Elizabeth* while Gardner and his party, including Syd Enever and Reg Jackson from Abingdon, followed later by BOAC airliner. The Bonneville trip was another success: sixteen international and American records. After the main efforts, a twelve-hour test on a showroom-stock MG TD Midget followed, with the car driven around the same course to underscore the fact that Gardner's car now sported such an engine, albeit of somewhat higher tune.

In tandem with the 1951 record-breaking effort, the Morris Motors Board was persuaded to support another MG friend—motor racer, writer, and photographer George Phillips—in his ambition to win a prize at the 1951 24 Hours of Le Mans. Having finished the 1950 event in his own MG Special, Phillips had the opportunity to secure the biennial Rudge-Whitworth Cup in 1951. So it was agreed that Abingdon would support his ambitions by building a Special based on the new TD Midget. The outcome, code-named EX172, looked unlike any previous roadgoing MG, but bore some resemblance to the still-current Jaguar XK120. Nobody yet knew it, but EX172 would form part of the inspiration

64

for the MGA that would follow barely four years later. Despite the effort made, the hopes, and the potential, Phillips managed to destroy the engine at La Sarthe in June; soon after EX172 was cut up.

Gardner was, by this time, laying plans for 1952. He expressed his desire to exceed his own two-liter records, with a Nuffield sourced engine instead of a Jaguar unit, plus other records with an evolved TD powertrain. The setup was similar to that of the previous year: the car crated and shipped by sea, the team flown out by airliner. The outcomes were largely successful, but with one unhappy upset. On one of the runs Gardner hit a marker post, which smashed the Perspex canopy and struck him on the head. Shaken but apparently unharmed, he carried on. Plans were subsequently made for a further speed session in Belgium in 1953, but when Gardner fell ill in the autumn of that year, his doctors advised him to rest. Sadly, Goldie Gardner never resumed his record-breaking adventures; he died in 1958.

ABOVE AND OPPOSITE: Just two of the three EX182 entries that made it through the 1955 Le Mans race. This is the car driven by Ted Lund and Swiss driver Hans Waeffler. They went on to complete 234 laps and finish in seventeenth place overall. Ken Miles and Johnny Lockett, in Car No. 41, finished in twelfth place, while the third car, No. 42, crashed in the sixth hour. The driver, Dick Jacobs, was injured in the accident, while his intended codriver, Irishman Joe Flynn, walked the circuit since he had missed his chance to drive it. *Author archive*

LAST OF THE OLD GUARD, FIRST OF THE NEW: MG *TF* MIDGET AND Z MAGNETTE

At the end of 1951, the Austin and Nuffield companies merged. Inevitably, the two businesses grew closer and began a process of rationalization—at first, principally of powertrains—but many aspects remained separate. What also became clear was who was in overall control: Len Lord. Six months

RIGHT: The creation of the modern, aerodynamic body style of the MG Series MGA was a bold and brave move. Many enthusiasts (including those at Abingdon) were keen on the design, and there was a major risk of alienating the hardened fans of the square-rigger T-Type Midgets, which the new generation was destined to sweep away. This is the cover of an early sales brochure, proclaiming the MGA as the "First of a new line." *Author archive*

BELOW: A major part of the MGA story was the brand-new chassis, designed by Enever and drawn up by his young assistant, Roy Brocklehurst. *Author archive*

later, Lord Nuffield, at 74 years of age, became the president of the new conglomerate, British Motor Company (BMC), and stepped away from any managerial role.

Gradually, the implications of this merger for MG became apparent. From a sports car perspective, the creation of the new Austin-Healey marque was a worry for Abingdon, and the gradual substitution of Austin engines for former Nuffield motors in the mainstream Morris and Wolseley lineup would clearly have an impact on production. Len Lord also relished his victory over Reggie Hanks, who had resisted the merger. Hanks stayed on as Nuffield chairman, but his power had been undermined.

MG was soon affected in two ways: First, MG's ambition for its own new sports car conception, code name EX175, was rejected by Len Lord, and the team were obliged to rework

the TD Midget into a face-lifted car, the TF Midget, still running with essentially the same 1,250 cc XPAG four-cylinder engine (a larger 1,466 cc XPEG would follow in 1954). Second, a project to replace the Y-Type with a new saloon designed by Gerald Palmer at Cowley was modified at a late stage in the design process to accept the new Austin-designed B-Series 1,498 cc engine in place of the XPAG. This car was the Z-Type Magnette, which used a new monocoque body.

Visitors to the October 1953 Earls Court Motor Show were treated to the conflicting spectacle of the "New, Air-smoothed Magnette" with its sleek Italianate styling, as well as the much more traditional-looking TF Midget; despite the changes, this Midget was clearly an evolution of the older generation of MG sports cars. Both were built at Abingdon, but the Magnette was very much a Cowley design (Cowley Drawing Office code DO1010) and the TF (EX177) unashamedly an Abingdon project.

The Magnette soon became a success story, and the original Z led to the later ZB in 1956 and Varitone variants. Meanwhile, the TF Midget would beget the TF 1500 when fitted with the larger XPEG engine.

CAPTAIN EYSTON AND CASTROL: A NEW ERA OF MG RECORD BREAKING

Goldie Gardner may have retired from record breaking, but it did not take long before George Eyston returned to MG. He was now a director at C.C. Wakefield, the makers of Castrol Oil, and he brought with him powerful sponsorship opportunities. Eyston had an American wife and spent a lot of time in the United States. As such he could see how British exports were fairing and understood the benefits of sporting success on the image of performance cars in particular. Since the end of World War II, Eyston had been courted by Len Lord, who at that time was only interested in Austin matters. The two had collaborated to publicize the new 1947 Austin A40, and a record effort in an Austin Atlantic at Indianapolis in April 1949. In 1954, there was the prospect of Eyston once more working with MG, now part of the BMC empire; it seems unlikely that any objection would come from Len Lord, at least.

ABOVE: The impact of the tragic events of the 1955 Le Mans race were far-reaching, although road rallying was deemed less risky than racing. Nancy Mitchell was one of the outstanding MGA drivers of her time. This photo shows just a few of the trophies she accumulated. *Bruce Chapman (Mitchell family)*

FOLLOWING PAGES: The jubilant MG team at the March 1956 12 Hours of Sebring, where all three cars finished, earning MG the honor of the team prize. Each car was prepared at a different garage although all were painted in the American racing color scheme of white with blue stripes. *Gus Ehrman*

Syd Enever was the key link between the postwar Gardner and Eyston eras. He knew that the older record car, which had been Gardner's personal property, had reached its practical limits in terms of pursuing higher speeds, and he managed to gain support and the necessary budget to develop a new record breaker, code name EX179. The car was developed using a spare chassis from the EX175 project. This was lightened with the time-honored Swiss-cheese approach of drilling hundreds of holes through the members. Terry Mitchell, at the time working at Cowley but later MG's chief chassis engineer, designed the body itself, which was produced by Midland Sheet Metal at Nuneaton. The finished racer was ready for its first trip to Utah in the summer of 1954. The principal driver was Ken Miles, a British ex-patriate living in California. (Miles is best remembered for his role in the Ford GT racing program a decade later).

EX179 played an ongoing role in MG's regular record-breaking efforts. A pause in 1955 allowed for an effort at Le Mans, and the car was reworked significantly in 1956, with a prototype of the forthcoming MGA Twin Cam engine fitted. This conversion was accompanied by a change from left- to right-hand drive due chiefly to the different exhaust stack arrangements for the newer

51
50

Designed to steal your heart!
The entirely NEW MG A
MG
White with black leather upholstery
is one of a rainbow of brilliant
new MG A color schemes available.
High in Style! High in Spirit!
There's promise of action in every line
of this sleek new beauty. And, promise becomes
reality when you experience the eager surge
of its powerful new engine—the sureness of big,
new oversize brakes—the solid road feel
that puts the new MG A in a class by itself.
Designed to steal your heart . . . make a
date for a test drive today!
Represented in the United States by
hambro
AUTOMOTIVE CORPORATION
27-29 WEST 57TH STREET
NEW YORK 19, N.Y.
Sold through a nationwide network of
distributors and dealers.
Product of
BMC

LEFT: The MGA featured in the November 1955 issue of the leading U.S. sports car magazine *Road & Track,* highlighting this BMC/Hambro advertisement on the back page (there was a photo of an MGA on the front cover too). *Road & Track*

BELOW LEFT: The major coup for MG with EX181 was the recruitment of British Formula One race driver Stirling Moss, who agreed to drive the car at Utah after meeting his other contractual race commitments for Ferrari at Monza. *Enever Family Archive*

BELOW RIGHT: Syd Enever and George Eyston stand on either side of the diminutive EX181 Roaring Raindrop at Wendover in August 1957. The stunning colors of this contemporary Kodachrome image show how the EX181 first appeared, in two tones of Metallic Blue for 1957's record-breaking efforts. The more familiar Green Livery only appeared in 1959. *Enever Family Archive*

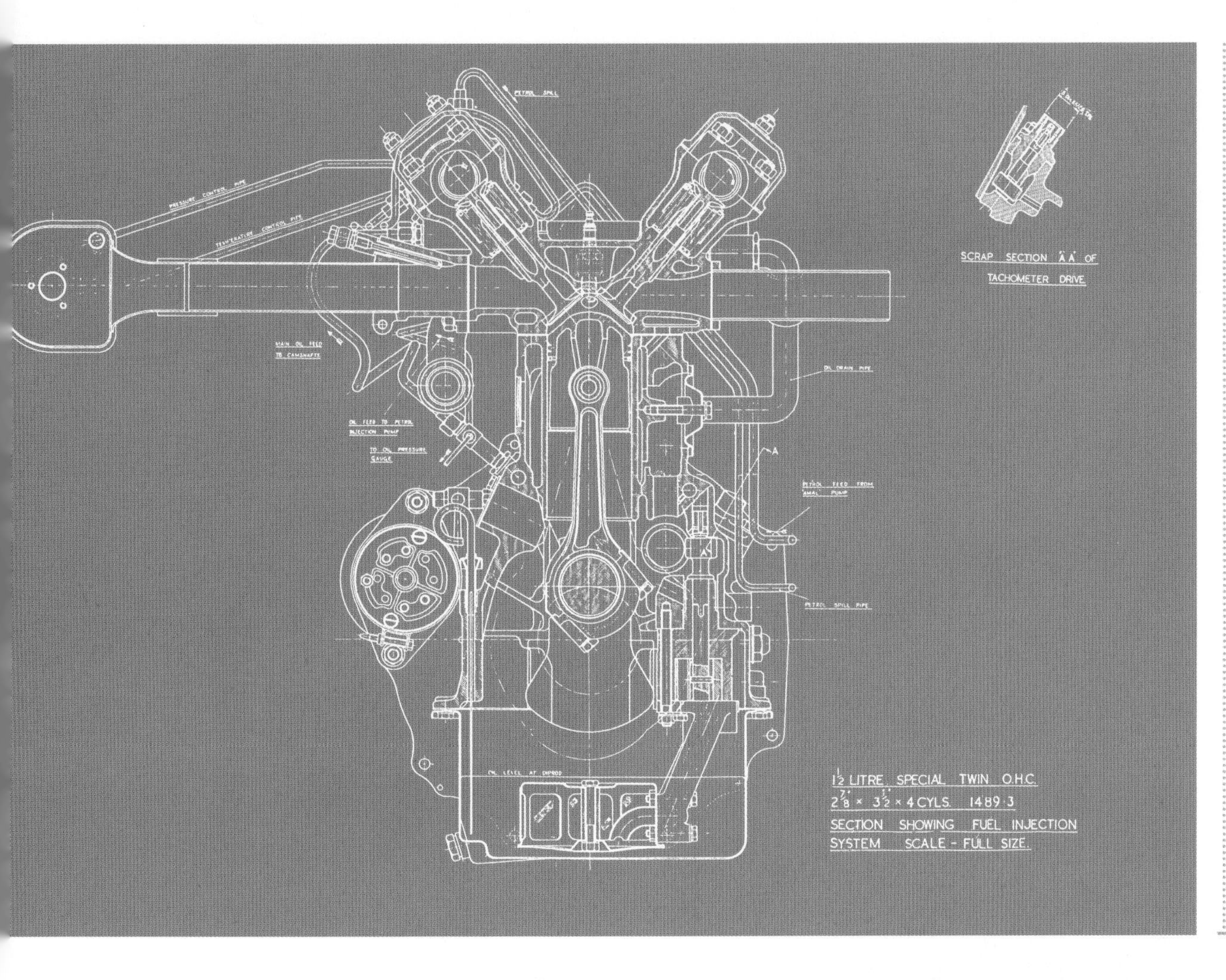

LEFT: The concept of a twin-cam version of the new BMC B-Series engine originally came from the fertile mind of Gerald Palmer. In addition to being a clever engineer, Palmer was also an accomplished engineering draughtsman. This is one of his drawings showing a cross section through his idea for a fuel-injected 1,498 cc B-Series twin-cam engine. *Gerald Palmer*

BELOW: EX181 streaks across the vast expanse of a dry salt lake in August 1957. Note the tail fin fitted at Eyston's insistence. *Enever Family Archive*

engine. In 1957, some months before the public launch of the Austin-Healey Sprite, EX179 was sent to Utah with a 948 cc BMC A-Series engine; its powertrain was diplomatically described as being "Morris Minor based." After another break for 1958, the car was rebadged as an Austin-Healey with a new project code, EX219. Though originally conceived as a completely new car, the exercise switched to a reworking of EX179 instead.

THE REVOLUTIONARY NEW MGA

Although Len Lord refused to sanction MG's new sports car when first presented to him, John Thornley did not give up. The case for change was further bolstered by feedback from the United States, where, despite the enormous affection for the traditional Midget style, there was a clamor for greater performance and more modern styling, especially in light of the new Austin-Healey from the same BMC stable. Thornley made it his mission to convince senior management of the viability of Abingdon as the BMC's natural home for a sports car center in terms of both design and manufacture. As a

ABOVE: Although the EX181 was undoubtedly the big event, the EX179 was still used alongside it. In 1957, it appeared as a "BMC experimental record-breaker with the 948 cc A-Series engine from the Morris Minor" (BMC publicity staff insisted that the record breaker should not be called a Morris Minor). Here we see Syd Enever (in the pith helmet he had acquired for his Utah visits) shaking hands with Dave Ash of Inskip, the New York MG distributor. *Enever Family Archive*

Success again!
Phil Hill gains six new
International Class E Records
Flying Mile 254·53 m.p.h.
FASTEST EVER
Driving an M.G. Special at Utah,
Phil Hill has established the following
six International Class E Records
(The engine is basically a B.M.C. 'B' series power unit
the M.G. Magnette, the M.G.A. 1600 and other B.M.C.
FLYING KILOMETRE . . 254·91
FLYING MILE . . . . . 254·53
FLYING 5 KILOMETRES . 232·97
5 MILES . . . . 238·36
10 KILOMETRES . . . . 234·49
10 MILES . . . . 191·03
In addition a number
American National
were
Safety Fast!

RIGHT: Naturally, BMC celebrated the success of the EX181, now with Phil Hill rather than Stirling Moss taking the honors. *James Mann*

consequence, a new design office was set up at Abingdon in June 1954.

Syd Enever recruited new staff, including Jim O'Neill and Terry Mitchell, who came over from Cowley. His first big task, beyond assembling his team and equipping the office, was to get on with the productionization of EX175. In this case, he would use the B-Series engine shared with the Z Magnette instead of the old Nuffield XPEG of the final TFs. The project code was still a Cowley one—DO1062—but it would be masterminded from Abingdon. O'Neill, with his body engineering experience, was key to the process of developing the new body shape. He worked closely with Morris Motors Bodies Branch at Coventry. Chassis work was largely down to Harry White, Terry Mitchell, and Syd Enever's protégé and heir apparent, Roy Brocklehurst.

John Thornley also had in mind a grand motorsports entrance for the new sports car, christening the MG Series MGA, with a series of lightweight versions for the 1955 Le Mans race. The MGA name was selected because, as Thornley explained, the alphabet had been used up, and by starting again at A (but including the marque initials in the model name), anyone talking about the car was more or less forced to mention the marque name when referring to it.

At the end of 1954, plans were laid to form a revitalized BMC Motorsports Committee, to be joined by new recruit Marcus Chambers, a seasoned campaigner. The first meeting of the new committee took place on January 13, 1955, and the minutes recorded prior progress with "the MGA 1500 already entered for Le Mans to replace the TF in June." It is interesting to note the early reference to the new MGA name and the confident expectation that the new model would be ready to launch in place of the TF1500 in June, around the same time as Le Mans. Intriguingly, the meeting notes further

The MGA Twin Cam was the easiest to spot, if you relied solely on your eyes and couldn't distinguish the different engine note. The center-lock Dunlop disc wheels were not that far removed from those used by contemporary racing Jaguars. Wire wheels were never offered on this model. *James Mann*

commented that a new-twin-cam engine was "supposed to be available in 1956." With that in mind, there was an optimistic suggestion of "an idea specification for the Magnette using twin-cam engine and disc brakes . . . with close ratio gears and knock-off wheels, with 300-mile fuel tank."

Unfortunately for the planned MGA launch that was to coincide with the Le Mans race, work on the production version of the DO1062 soon showed problems with the body tooling, which meant unavoidable launch delays. Work on the lightweight Le Mans cars continued under the Abingdon EX182 project code. Following the production delays, Thornley negotiated with the ACO, the Le Mans organizers, for the MG entries to be regarded as prototypes of a "possible new sports car." Hence, a team of beautiful British Racing Green EX182s departed in June 1955 for that year's event.

There were four cars built from March 1955 onward, with shakedown testing at Silverstone the following month. Initial testing was undertaken by the development mechanics, led by Alec Hounslow. Circuit assessments were initially by Dick Jacobs, who later wrote how, at a second session,

ABOVE: Nuffield Exports had a massive warehouse where around three hundred MGAs per week could be stored. This photo, taken in 1957, shows vehicles slated for shipment to Los Angeles, where they would be received by Gough Industries, the local West Coast distributor. ***Author archive***

RIGHT: The special MGA Twin Cam owned by Ted Lund survives in the care of present owner Steve Dixon. ***Author archive***

THE EX 182 LE MANS CARS								
Chassis Number	**Engine Number**	**Le Mans 1955 Race Number Prior to Scrutineering**	**Le Mans 1955 Actual Race Number**	**First Recorded Registration Number**	**Mechanics Driving to the Race**	**Drivers in the Race**	**Finish Overall**	**Finish in Class**
EX 182/ 38	EX 182/ 43	40	41	LBL 301	Alec Hounslow	Lockett/ Miles	12th	5th
EX 182/ 39	EX 182/ 42	41	42	LBL 302	Jim Cox	Jacobs/ Flynn	DNF	DNF
EX 182/ 40	EX 182/ 44	64	64	LBL 303	Cliff Bray	Lund/ Waeffler	17th	6th
EX 182/ 41	EX 182/ 45	Test car	Test car	LBL 304	Doug Watts	Practice car	N/A	N/A

"we covered some 80 to 90 laps, being well over 250 miles [402 kilometers] during the day." He added that, thanks to changes made following his first session the previous week, "it was now possible to drift the fast corners in the approved style and the car felt right, which gave the drivers more confidence. . . . there was still some engine fume trouble but it was not as bad as the previous tests. The new positions for the seat, pedals and steering wheel made the car much more comfortable. We came away from that test day much happier at the progress that was being made." Further testing in April brought Ken Wharton and motorcycle racer Johnny Lockett to try the cars. After these test sessions, Marcus Chambers's report to management concluded that "performance tests at Silverstone had shown that a sustained lap speed of 85 mph, with a max lap speed of 94 mph, was theoretically possible."

The cars were finished and driven in convoy to the Burlington Hotel in Folkestone, along with the elegant BMC Competitions truck—or *camion*, as John Thornley dubbed it in honor of its initial function in supporting the trip to France. The party crossed the channel to Boulogne-Sur-Mer on the *Lord Warden* ferry. Once in France, the group drove south, pausing for lunch at Quai de la Canche before convening at Château Chêne de Cœur, home of La Comtesse de Vautibault, a friend of George Eyston.

The race itself, with three EX182s competing (race numbers 41, 42, and 64), was most memorable for the horrific fatal crash involving the Mercedes-Benz of Pierre Levegh and the Austin-Healey 100 of Lance Macklin. Unfortunately, the MG team did not survive unscathed, as Dick Jacobs crashed his EX182 (No. 42) at Maison Blanc corner. The car turned over and burst into flames, and Jacobs was rushed to the hospital at a time when emergency services were fighting to deal with the multiple fatalities and serious injuries from the Levegh/Macklin crash.

Ted Lund in No. 64 struck a stationary Jaguar D-Type, which had been abandoned on the circuit, but after repairs in the next pit stop the car was able to continue with Hans Waeffler at the wheel. Both the surviving cars made it to the end.

After the race, Dick Jacobs made a slow but steady recovery. He remained active in MG racing, but it was as owner and manager rather than driver. The popular press made much of the Le Mans fatalities, and for a while it seemed that European motor racing would

The EX186 was one of those projects created under the radar. Once BMC management became aware of its existence, the decision was made for it to "disappear." Fortunately it was not destroyed: The vehicle was spirited away to San Francisco, into the care of Kjell Qvale, and later rediscovered and lovingly restored by enthusiasts Joe and Cathy Gunderson. *Joe and Cathy Gunderson*

ABOVE: Rarer even than the MGA Twin Cam was the MGA 1600 Deluxe (not its official title), which was created to "use up" the surplus Twin Cam chassis with, initially, the earlier 1600 engine and then the 1,622 cc Mark II engine. This particular specimen belongs to enthusiast David Halliday and is one of those originally advertised for sale in *Autosport*. *David Halliday*

suffer accordingly. Although MGs were not involved, a subsequent crash in September at the Ulster TT race at Dundrod around the time of the international launch of the MGA brought a change in racing policy for BMC. George Harriman, Len Lord's deputy, issued the directive that BMC would henceforth withdraw from races, although they would remain active in rallies such as the Mille Miglia, which was arguably more a road race than a true rally. The immediate consequence was that any return to Le Mans in 1956 by MG, at least as a factory team, was ruled out along with tentative plans for a new bespoke MG Le Mans racer.

THE MGA LAUNCH

The world debut of the MGA was staged at the Frankfurt Motor Show. The red show car traveled in a special aerodynamic pod, the Speedpak, stowed under the fuselage of a Lockheed Constellation, which flew the car to Frankfurt's airport. There it was met by the local agents, carefully unpacked, and driven to the show in time for the opening.

It was a spectacular piece of theater, and the new MG was next destined to appear at the Paris and Earls Court shows. In many ways the MGA was a bold move: Its aerodynamic body transformed top-end performance and won new converts, but there were disgruntled traditionalists who were less happy about the changes.

Sales over the following seven years of well over one-hundred-thousand units proved the merits of the change. At first the engine was the still new 1,489 cc B-Series, but from May 1959 the capacity was raised to 1,588 cc, matching that of the twin cam launched the previous summer.

RACING AND RALLYING THE MGA

Fortunately for MG, the motorsports directives in Europe did not translate to North America's

racing scene. In 1956, a team of three MGAs raced at the 12 Hours of Sebring with oversight from Hambro Automotive. The cars were finished in American racing white colors but were prepared at different shops in New York, Washington, D.C., and Warren, Pennsylvania. All three MGAs finished (fourth, fifth, and sixth in their class) and won the team prize. The experiment was repeated the following year, and again all three cars finished—first, second, and fourth in their class this time, with another team prize into the bargain. There were no MGAs at Sebring in 1958, but they were back in various forms from 1959–1962. Every car ever entered finished, adding to MG's prestige.

The MGAs also performed well in European rallying, with appearances at the Monte Carlo, Alpine, and Tulip rallies, and memorably at the Mille Miglia road race in its final years. Among the highlights were stunning performances like that of Nancy Mitchell in the big European rallies; road races like the Mille Miglia, Alpine Rally, Liège-Rome-Liège, and Lyon-Charbonnières Rally; and other MGA races at the likes of Oulton Park and Nürburgring by aristocratic privateers Richard "Fitz" Wentworth-Fitzwilliam and Robin Carnegie.

ROARING RAINDROP AND TWIN CAMS

Notwithstanding the creation of EX179 as a brand-new record breaker, Syd Enever realized from the outset that its structure and form were barely less limiting than that of the older Gardner car. Accordingly, Terry Mitchell—now ensconced at Abingdon—was tasked with taking Joukowski profiles (aerofoil sections) and developing a new, bespoke teardrop-shaped record car (EX181), which was soon dubbed the Roaring Raindrop when it appeared in the summer of 1957. EX181 was revolutionary in that the twin-Shorrock supercharged twin-cam engine was mounted amidships, driving the rear wheels through a converted MG TC gearbox. Rear suspension was of the De Dion type, and the driver was seated toward the front of the car, his feet barely inches from the tip of the nose.

EX181's Bonneville debut was in August 1957 at the hands of famous British Grand Prix driver Stirling Moss, who achieved a two-way average speed of 245.64 miles per hour (395.3 kilometers per hour), breaking five International Class F records. The basic engine, in common with that of the 1956 iteration of EX179, was a twin-overhead-cam version of the MGA's B-Series, yet to appear

ABOVE: The MGA 1600 Mark II was available in either coupe guise or as the usual roadster. *Jonathan Stein*

OPPOSITE BOTTOM: The first significant development of the basic MGA (other than the Twin Cam) was the introduction of the MGA 1600, with an engine capacity that was unique to the model. This photograph was, appropriately, taken at Abingdon Marina. *James Mann*

The MGA 1600 Mark II's most obvious visual difference was the indented radiator grille. Meanwhile the tail lamps at the other end were effectively ADO15 Mini units mounted horizontally on small plinths. *Author archive*

in production form. That would change in the summer of 1958 when the definitive MG MGA Twin Cam was presented to the media at the FVRE test track at Chobham.

In its short life, the MGA Twin Cam was both the apotheosis of postwar MG sports car development and nearly the end of MG's independence. Its specialist and highly strung nature meant that it too often fell afoul of abusive owners and their mechanics. This, coupled with a number of fundamental faults that were only fully understood and addressed after the model had left production, cemented its demise only two years after it debuted. Rightly regarded today as one of the more desirable Abingdon MGs, it is also one of the rarest, with just 2,111 production cars produced.

Apart from Chobham, appearances of EX181 in 1958 were confined to low-speed demonstration runs at Silverstone with Syd Enever or Alec Hounslow on hand. For 1959, however, the record car returned to Utah for what would prove to be its final and greatest speeds. This time Moss was not available, so the honor of pursuing new records fell to the great American driver Phil Hill, who took the Roaring Raindrop to 257 miles per hour (414 kilometers per hour). Along with EX181 was EX219, which was actually EX179 rebadged as an Austin-Healey for marketing purposes.

Although Enever fostered ambitions to take the record even higher using an all-new low-slung car (EX233) based on a supercharged, inclined transverse 948 cc A-Series powertrain, circumstances were against him. In late 1961, Alec Issigonis, being lionized on the back of his recent small car prowess, became BMC's technical director following the retirement of Nuffield man Sydney Smith. Almost immediately, the shutters came down on projects with which Issigonis had no direct involvement—more particularly if they involved the use of his creations. EX233 got as far as a scale model and some paper studies, but no further—a fate not dissimilar to the later Mini-based MG sports car exercises dreamed up variously at Longbridge and Abingdon (ADO34/35/36), all of which were eventually sidelined.

TED LUND AND LE MANS

Although the factory's ambitions to return to Le Mans in 1956 were curtailed by the racing

ban, this did not prevent behind-the-scenes support for a private entry under the auspices of Ted Lund, one of the 1955 EX182 drivers and a member of the MG Car Club's North-West Centre. The team focused their efforts around a hybrid, lightweight MGA Twin Cam built with the benefit of donor parts left over from the factory's EX182 efforts. Registered as SRX210, the car was driven by Lund at Le Mans from 1959 to 1961, evolving with each year. In its last race in France, it sported a bespoke coupe body, largely the work of MG body draughtsman Don Hayter and given the project code EX212.

THE MGA AND MAGNETTES EVOLVE

With the MG Z Magnette, then the MGA Roadster and subsequent Coupé on offer, the next steps were to fine-tune the product lines. As we saw earlier, this first saw an increase in engine capacity for the MGA, the MGA 1600, with the same engine capacity as the twin cam (1,588 cc) but without the mechanical complexity of that model.

When the MGA Twin Cam was canceled in the spring of 1960, a model was created to use up the remaining Special chassis. The MGA De Luxe married the normal B-Series in 1600 guise (both Mark I and Mark II De Luxe models were built). The MGA De Luxes were hardly publicized:

TED LUND, LE MANS, AND SRX 210						
Year	**Days**	**Body Configuration**	**Color(s)**	**Drivers**	**Car No.**	**Notes**
1959	June 21–22	Roadster "EX 182" body	Ash Green with British Racing Green sides	TedLund/ Colin Escott	33	Collision with Alsatian dog in the 21st hour of the race (on the Mulsanne Straight) led to overheating and retirement from the race.
1960	June 25–26	Coup with conventional grille	Metallic Green all over ('EX' Green)	Ted Lund/ Colin Escott	32	1,762 cc and twin Weber 40 DCOE carburetors. Finished twelfth on distance and won two -liter class.
1961	June 10–11	Coup now with streamlined nose	Metallic Green with BRG side flash	Ted Lund/ Bob Olthoff	58	Another 1,762 cc engine; twin Weber 45 DCOE carburetors. Car retired after fifteen laps with engine trouble.

They were an expedient way of using up those special chassis with their center-lock disc wheels and four-wheel disc brakes, partly to avoid embarrassment of suppliers like Dunlop.

The first batch of twelve was advertised by MG's London distributor, University Motors, in the July 29, 1960, issue of *Autosport*. Today, these cars hold a mythical status, almost as highly prized by collectors as genuine MGA Twin Cams.

The Magnette evolved into the ZB and the later Varitone version, the latter available with two-tone paint schemes and featuring an enlarged rear window. In 1959, however, the much-loved Z Magnette was replaced by an MG member of the new BMC Farina saloon range and badged the MG Magnette Mark III. This new model retained the same 1,498 cc B-Series engine, but the platform was new and unwieldy. The MGA 1600 Mark II saw the introduction of a much-improved 1,622 cc version of the B-Series engine that, unlike the earlier 1600, was shared with other BMC models.

The MG Magnette Mark III was one of the new Farina BMC range. This example was photographed in Canada in 1960. *Author archive*

230

5

Superlative Sixties

THE MIDGET RETURNS

The arrival of the Austin-Healey Sprite was an opportunity for the Abingdon team. To a great extent a cuckoo of the Healey family's creation, it provided a neat entry-level sports car that usefully expanded the BMC sports car range and helped keep the Abingdon lines busy. John Thornley and Syd Enever had been exploring ideas for such a car for a number of years, prior to the Sprite's arrival in May 1958. Now it was obvious that senior management would not sanction a second baby sports car, and so the next best solution was to develop an MG version of the Sprite—with the Midget name, of course.

The project to create this new Midget, one featuring a sub-one-liter engine for the first time in decades, was twin-tracked with a face-lift for the Sprite. The key requirement for the latter was to reshape its controversial front-end styling.

Initially the two projects were separate (ADO41 was the Sprite face-lift, ADO47 the new Midget), but production became closely aligned following behind-the-scenes communications between Abingdon and Warwick. Largely identical bodywork was applied to both cars, such as more conventional front and rear wing lines, and a proper boot with opening lid. Although Enever and O'Neill were the key body design people, responsible for the new ADO41/47, it was body draughtsman Denis Williams who took on the bulk of the actual design work. Williams told the author that "the MG Midget Mark I and Sprite Mark II were developed using the Sprite's original

LEFT: A works MGB ("Old Faithful," MGB Roadster GRX307D) and one of the seminal Mini Cooper S entries (Paddy Hopkirk and Henry Liddon's car) in the course of preparation for the 1966 Monte Carlo Rally. Tony Fall and Ron Crenin drove the MGB but retired; the controversy of the disputed Mini results dominated the news after the event. *Author archive*

The Farina Magnettes were seldom as popular as their predecessors, so fewer have survived. This is a 1966 MG Magnette Mark IV at an MG Car Club event, parked alongside the more familiar rounded shape of the MG ZB. *Author archive*

underframe as a basis, having a completely new, restyled front end and rear end, the scuttle and doors remaining unchanged. The new rear-end, full-size layout was drawn by me at Abingdon, and the front end in collaboration with Pressed Steel and Swindon, except for the radiator grilles and surround, which were designed at MG."

The ADO47 Midget was the first MG to benefit from the new plastic, shield-shaped grille badge, appearing a year before the launch of the MGB. Similar to that car's legacy, it is entirely wrong to claim, as others have, that Denis Williams simply borrowed the rear wing line of the MGB for his new ADO41/47 pair. All this design work was ultimately directed by the chief body engineer Jim O'Neill.

The new small sports cars were unveiled to the public in May and June 1961. In typical BMC style, there were separate unveilings for the Austin and Nuffield families of dealers and enthusiasts. The Mark II Sprite was pitched as a slightly cheaper model, whereas, with a little more chrome trim, slightly fancier interior, and the traditional slatted, plated grille, the MG Midget was priced a little higher. Both had the same twin-carburetor 948cc A-Series engine, and their performance was to all intents and purposes identical. Through 1961 and into the following year, the new Midget was sold alongside the MGA—the latter in its final, 1600 Mark II guise, while the family man could opt for the larger MG Magnette in its new Farina form. (The 1,489 cc ADO9 MG Magnette Mark III took over from the ZB Magnette in 1959, and this was the model that evolved into the slightly better ADO38 Mark IV in 1961, the latter sharing the bigger 1,622 cc engine of the final MGA Mark II.)

The Midget and Sprite, known collectively to many as Spridgets, evolved throughout the 1960s and beyond into the following decade. The model sequence was always one adrift between the two: The Sprite Mark III came with the Midget Mark II, and the Sprite Mark IV aligned with the Midget Mark III. Bigger engines (1,098 cc, and then 1,275 cc), new rear suspension (conventional semi-elliptic springs superseded the original quarter-elliptics), modest improvements in trim, and the refinement of a superior hood top and winding side windows with hinged front quarterlights (window vents) all added to the appeal of BMC's smallest sports car, although it was always rather overshadowed by what arrived in 1962 to replace the MGA.

HYDROLASTIC SUSPENSION COMES TO MG: ADO16 MG 1100

From a BMC perspective, the start of the new decade was arguably dominated by the excitement and popular success of the new ADO15 Mini, initially in the guises of the Austin Se7en and Morris Mini Minor (or simply 850 in some export markets). A sporting MG-badged offshoot was contemplated, but it was soon rendered unnecessary by the emergence in 1961 of the new Cooper variants. Above the Mini there was clearly space in the range for a larger, family-oriented saloon; this duly arrived, initially as a Morris 1100, in 1962.

Because the Austin franchise benefited from the crisply styled Austin A40 saloon, which was rather more modern in style than Morris's veteran Minor, the Nuffield franchise was the first to take advantage of this new front-wheel-drive ADO16 range. The Morris 1100 was soon followed by an MG version, complete with the distinctive grille shape and, for export only, a sleek two-door version with raked rear-door window pillars that lent it a certain coupe style. All variants employed Alex Moulton's revolutionary fluid-based Hydrolastic suspension, which offered an unrivaled (if sometimes rather floaty) ride quality, often likened to that of much larger cars.

The ADO16 may have been largely forgotten by the public, but in its time it was recognized as a remarkable product, with a combination of packaging genius and crisp Pininfarina-detailed styling. It became the template for a number of later European rivals and undoubtedly helped inspire the 1970s and 1980s generations of five-door family hatchbacks. Its lack of a tailgate (on the saloon versions) was one feature derived from the market's continuing preference in 1962 for a separate boot. The BMC 1100 range was, for a time, the U.K.'s best-selling family car, expanding domestically to embrace the Austin, Riley, Wolseley, and Vanden Plas marques as well as the original Morris and MG variants. The 1100 was only later eclipsed by Ford's Cortina.

The MG Midget prototype at Abingdon. *Enever Family Archive*

ABOVE: The lucky owner of this beautiful Tartan Red MG Midget Mark I 948 is Perry Massey of Paso Robles, California. Notice the rare wheel trims. *Perry Massey*

RIGHT: Initially 948 cc and, from late 1962, 1,098 cc, the engine of the Midget was the redoubtable A-Series four cylinder, already seen in the Austin A35, Morris Minor 1000, and—in transverse-mounted front-wheel-drive form—the revolutionary new ADO15 Mini. Nevertheless, Abingdon's smallest sports car retained a traditional inline drivetrain layout. *Author archive*

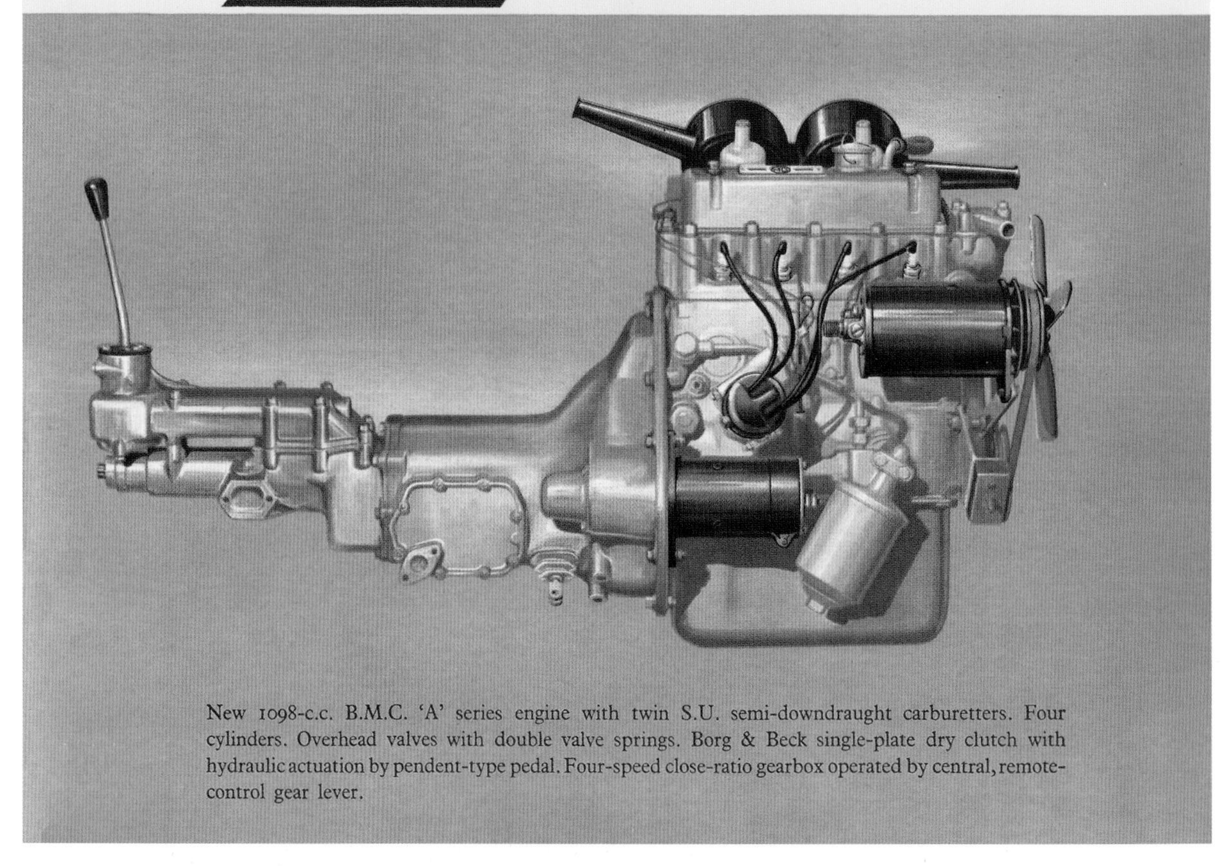

Starts ahead
New 1098-c.c. B.M.C. 'A' series engine with twin S.U. semi-downdraught carburetters. Four cylinders. Overhead valves with double valve springs. Borg & Beck single-plate dry clutch with hydraulic actuation by pendent-type pedal. Four-speed close-ratio gearbox operated by central, remote-control gear lever.

In 1967, a larger engine was offered as an optional extra on the 1100 (1,275 cc), and this was soon upgraded to the MG 1300, whose sales continued in the U.K. until 1971. Few know that the Sport Sedan might have staged a comeback, perhaps in the form of a 1500 E-Series (as it actually *did* appear in Australia), but any plans for an MG 1500 were ultimately abandoned. At its peak, around a thousand ADO16s were made every day through the key plants at Cowley and Longbridge. Exports covered a number of markets over the years; local assembly was undertaken in Italy by Innocenti—who also built and sold a special sports car based on the Sprite underframe—in Seneffe, Belgium, and in more modest volumes in other markets such as Ireland.

In the United States, there were great hopes for what was branded there as the MG Sport Sedan, offered as the intelligent driver's alternative to the archaic and primitive (but ruggedly reliable) VW Beetle. Considerable marketing dollars were expended on clever advertising, with Sport Sedan adverts appearing in glossy publications like *National Geographic*, but the car itself failed to meet sales expectations in the United States, where customers bought it in honorable but never earth-shattering numbers.

Reliability woes eventually undermined the Sport Sedan. BMC tried again in 1967 with a face-lifted version called the Austin America, featuring a new automatic transmission, but that is a story for another book.

ABOVE: The ADO 34 proposal was a neat idea: to make a kind of Mini Cooper sports roadster. It was developed in two forms, one by Abingdon and the other at Longbridge with Pininfarina input, seen here. It ran afoul of corporate politics and competing priorities. *Author archive*

The New MG MIDGET MARK II SPORTS CONVERTIBLE
• NEW SUSPENSION
• INCREASE IN POWER AND SMOOTHNESS
• NEW DE-LUXE COCKPIT

LEFT: The MG Midget Mark II followed early in the MG Midget story. A contemporary of the Austin-Healey Sprite Mark III, some of the more obvious evolutionary changes in this car were quarter lights (that is, vent windows in the doors), a change to semi-elliptic rear suspension, and a larger engine: at first 1,098 cc, later, with the Mark III of 1966, 1,275 cc. *Author archive*

THE SUPERLATIVE MGB

Aesthetically, the MGA represented a remarkable change of approach from the traditional styling of the cars it replaced, but under the skin it was much the same as what had gone before, especially where it used a separate chassis and body and relied on fairly primitive wet-weather equipment. Thoughts of a replacement began even as the MGA entered production, and the initial expectation was that the next generation would be little more than another rebody: However, before long, this changed to a realization that the next model would instead use modern and lighter, yet stiff, monocoque construction. This philosophy was exemplified by the Z Magnette, as well as in a design concept for a sports car known as the MG Magna by Gerald Palmer, and in other vehicles, including the Austin-Healey Sprite. What they saw in these new vehicles encouraged John Thornley and Syd Enever to shift their allegiance.

As Thornley once told the author, "In November of 1955, Sydney Enever and I produced a joint paper, 'Suggested Design and Development Programme for Abingdon Products,' in which we argued quite strongly that all future Abingdon cars should have chassis frames. Yet, by the middle of the following year, we

LEFT: A lot of hope was riding on the MG 1100, marketed in the United States and a few other countries as the MG Sports Sedan. After all, it was dynamically modern, offering an excellent ride, handling, and packaging—all attributes that were vastly superior to the market-leading VW Beetle. Unfortunately, durability and other factors couldn't match the German opposition. *Author archive*

TODAY'S SMART CAR IS THE
MG
Princess 1100

RIGHT: The MG Princess 1100 was rather eccentric—and short-lived. A version of the Vanden Plas Princess 1100 sold elsewhere, but in the United States it was marketed as a kind of MG mini–Rolls Royce. It was expensive and found few takers, much like the contemporary Vanden Plas Princess Four-Litre R, which was briefly offered in North America. *Author archive*

were scratching out the beginnings of the B, which gradually became monocoque!"

Designs for what would eventually become the definitive MGB began with EX205, a car with a separate chassis. By 1958, these evolutions led to a new chassis-less sports car coded EX214. Once this project received Morris Motors Board approval that year, it was given the corporate project code ADO23. The layout was determined by Enever—"the father of the MGB." as Thornley unequivocally called him—with the chassis design work largely down to Roy Brocklehurst, Terry Mitchell, and Richard "Dickie" Wright. The body and trim design was overseen by Jim O'Neill, but the detail work was largely a joint venture between Don Hayter, Don Butler, and Eric Carter of Morris Motors Body Branch, Carter having had a key design role in the MGA as well as in this new model.

Early thoughts of an all-independent rear suspension for this car were soon discarded, not least because, other than on the Austin Gipsy 4x4, there was no suitable rear differential setup, which meant a shift to a coil-sprung live axle. Enever and some of his chassis team had serious doubts upon testing this design, and so, at a late stage, the plan switched to the tried-and-tested semi-elliptic rear-suspension approach, resulting in the need for Don Hayter to revise the outer skin lines of the back end of the bodywork to provide the definitive MGB line.

At the front end, there was much debate about the form of the front wings and grille, with initial thoughts of a conventional headlamp mounting and a rectangular grille, both broadly modeled on the style of the one-off Frua-built EX205/2 MGA prototype, which had been constructed at the behest of BMC deputy manager George Harriman.

It was Enever, however, in conjunction with Jim O'Neill, who jointly alighted on the idea of pocketed headlamps. These were like those of the 1956 Ferrari 250GT California, which were also seen on the Frua-designed Renault Floride

The MG1100 was the second variant of the ADO16 BMC 1100 range launched, shortly after the Morris 1100 and alongside the new MGB. This example, although bearing U.K. plates, is the export-only left-hand-drive two-door version, its slightly longer doors and raked B-door pillar giving it something of a coupe style. *Author archive*

ABOVE: The first serious move toward replacing the MGA had been the commissioning of a one-off prototype on an MGA chassis by Frua of Italy. This 1957 Kodachrome photo of the car comes from Syd Enever's personal archive. Although much respected, the Frua was perhaps too Italian and upmarket for a 1,500 cc English sports car. It was measured and drawn up (as EX205/2), a model was made (which survives), but the beautiful, expensive, hand-built prototype was cut up to avoid paying import duty. *Enever Family Archive*

LEFT: This model was well on the way toward the definitive MGB style. Note the MGA-like rectangular grille—Enever later decided to go for the broader chrome grille seen in production—though the headlamps are already set slightly back into the wings. Silver side trim is already in place, ready should Enever's preferred duotone paint schemes ever be offered (they never were in production). The rear wing line has a slightly more pronounced fin shape than the form chosen for production. Immediately behind this EX214 model is the Harry Herring model, made as a copy of the Frua MGA. *Author archive*

This neat scale model of the future MGB—EX214 or ADO23—was made in wood by Abingdon's resident model maker, Harry Herring. Here we see the definitive lines emerging—the characteristic broad grille and clean wing lines—undertaken largely by Don Hayter under the direction of Enever and his body design deputy, Jim O'Neill. *Don Hayter*

that had first appeared at Paris in 1958 and then the next spring in Geneva. Following a visit to the Geneva show, Enever sketched out what he wanted, with a slimmer, broader grille shape, and the detail work was duly undertaken to great effect by Don Hayter. In fact, Hayter deserves much of the credit for the final style of the MGB.

Renault tried to claim that BMC had copied their design, but they soon abandoned their complaint. Nevertheless, from then on, Thornley and Enever generally held publicly to the story that the design of the MGB had originated in that of the EX181 Roaring Raindrop, a concept that O'Neill often laughed off as stretching credulity. Another popular rumor was that Pininfarina refined the design of the MGB, which is also nonsense: Credit should be fairly shared by Enever, O'Neill, and Hayter, the latter doing the lion's share of the actual work, albeit under the direction of the others.

Production of the new MGB started in the summer of 1962, allowing it to overlap with the last of the MGA Mark II models. These included a hundred-thousandth Special, painted in gold and upholstered in white, which was exhibited at the April 1962 New York Motor Show. By the autumn of 1962, the imminent arrival of a new MG sports car was something of an open secret, with preliminary specification sales and marketing literature going to trusted parties from July onward. The press reveal took place in September, there was an unveiling at University Motors in London, and the MGB even made a minor appearance at the Paris Salon. The big splash, though, was to be at the Earls Court Motor Show in October, where the MGB shared BMC honors with the Morris and MG 1100, a larger-engined (1,098 cc) MG Midget. Elsewhere at the show, visitors could see the new Triumph Spitfire and the more prosaic but highly significant Ford Consul Cortina in its original 1200 guise.

At the launch, Enever stated that "our aim is to give Sports car motoring to as many people as possible, and to give it to them at the right price. We do not want to make a small quantity of high-priced, specialized cars for the few. At the same time, we set out to provide the fastest possible car, combined with the greatest possible degree of safety. The MGB was developed with these intentions, and in this car, although the ride is much softer, we have good controllability and with no vices under extreme conditions." After describing many

RIGHT: MGB customers got their first clue of the new MG sports car's look in a limited-circulation *Advanced Specification* brochure. It was published from June 1962 on and sent to dealers to entice regular customers or new prospects. *Author archive*

Maintaining the breed with streamlined Power
THE NEW
MG
SERIES MGB
WITH 1800 c.c. ENGINE
Safety fast!
PRELIMINARY SPECIFICATION—CONFIDENTIAL

BELOW: The dashboard of a 1964 home-market MGB. The three-spoke steering wheel continued an MG tradition: Chief designer Syd Enever liked large steering wheels. *Author archive*

of the car's salient features, he concluded, somewhat presciently, "as you know, the MGA was a highly successful car, and the MGB should be more so."

Introduced as a 1963 model, the MGB had begun to be shipped, road-tested, and sold in the United States from late 1962, and it was driven in a demonstration session at the San Carlos Ranch by notable drivers such as Juan Manuel Fangio. The major sales effort in the American market really took off in the spring of 1963, however, with a show debut at New York in April. A month earlier, two MGBs had been entered at that year's 12 Hours of Sebring, but neither finished. Blame was laid on testing problems back home during the freezing 1962–1963 U.K. winter, but the lack of success after a solid track record at the circuit with the MGA was a source of some embarrassment.

In any event, the MGB was well received by the world's press. Its enlarged engine, at 1,798 cc, but still initially with three bearings and no bottom gear synchromesh, met with general approval, as did its much-improved creature comforts, such as winding door glass, hinged front-quarter glass, generous cockpit capacity, and the vehicle's clean new styling. With the full MG range now comprising Midget, MGB, MG1100, and MG Magnette, the marque now offered the fullest variety of models since the heady prewar days of Cecil Kimber.

Within just a few years the variety would grow even more, with the arrival of first the MGB GT and then the MGC range. For a brief period around the end of 1967, there would be the neat number of eight variants available (if that year's MG1300 was counted as a distinct model).

MGB GT: THE POOR MAN'S ASTON MARTIN

The coupe had usefully broadened the MGA range, but it had always sold in low proportion to the open equivalents, and its style was less admired because it resembled the optional bolt-on hardtop. For the MGB, Thornley and Enever had in mind something more distinctive. There were a number of pointers toward a new style,

The Passing of MG's Original Patron

The final curtain for Lord Nuffield, the man who started this whole story, came on August 22, 1963. In his later years, especially following the BMC merger of 1952, he focused more on his celebrated philanthropy in the world of medicine. While he retained little influence in terms of the car business, he nevertheless continued to come into his office at Cowley on a regular basis. He had no children, and his wife predeceased him. Their home, Nuffield Place, is now maintained by Britain's National Trust, and is open to the public.

The first production left-hand-drive MGB (GHN3-L-102) survives in private hands in the United States. The Iris Blue paint color is a carryover from the MGA 1600 Mark II, which was used extensively for U.K. and U.S. launch publicity. Being an early U.S.-specification car, the lenses at the front are all white rather than the amber/white seen in most other markets. (Notably, Italy also took white front indicators.) *Shad Huntley*

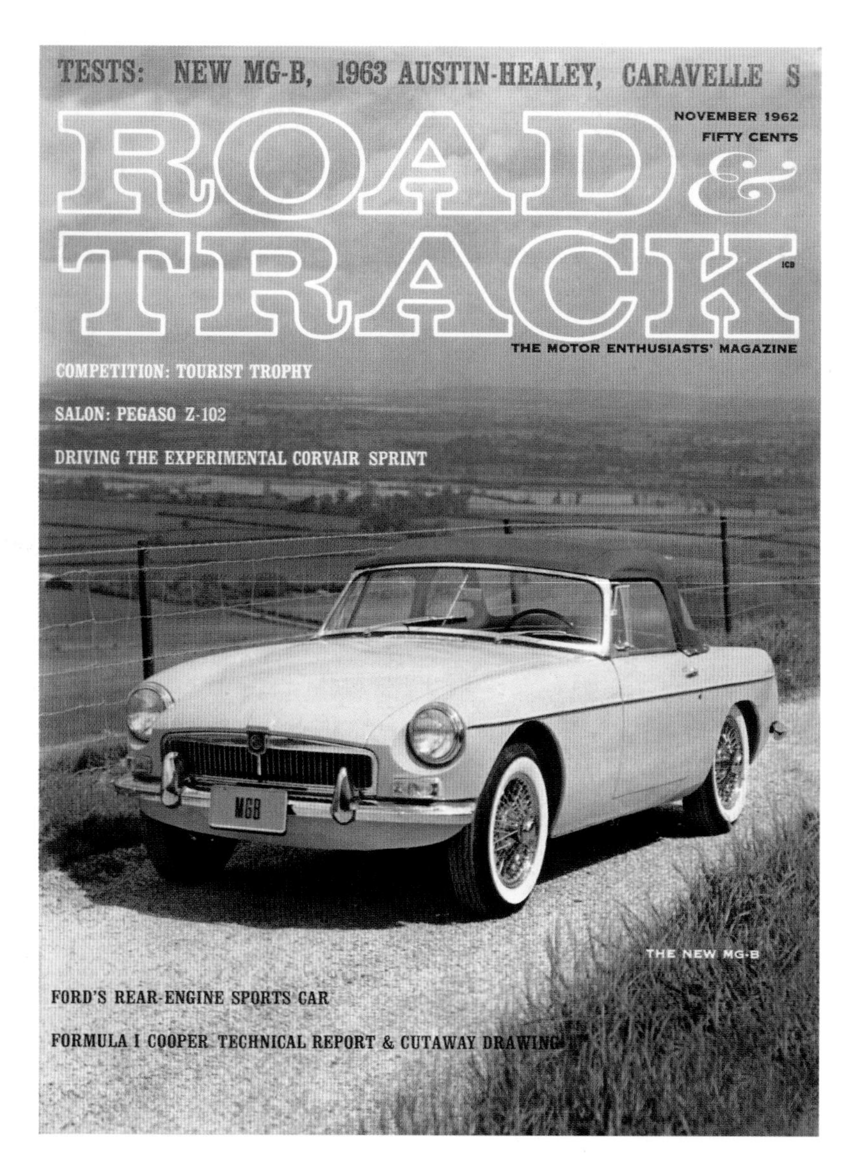

ABOVE LEFT: Unsurprisingly, the new MGB made the cover of *Road & Track*. *Road & Track*

ABOVE RIGHT: In MG's home market, the new sports car was described as the "Superlative MGB." In the United States, early advertising proclaimed that "everything is new but the octagon." *Author archive*

RIGHT: A sign of the changing auto sales landscape in the 1960s: emissions testing and, as seen here, crash testing. This MGB appears to have fared quite well in its test at the Motor Industry Research Association test facility. Roy Brocklehurst of MG stands at far left. *Author archive*

including the shape of some Aston Martins that Thornley had admired in a race at Silverstone. Further, when the Jacobs Midgets were built in 1962 with alloy coupe bodywork, their shape clearly indicated a direction for something similar in the larger MGB.

By 1963, Syd Enever and body draughtsman Jim Stimson had drawn up a possible conventional fastback style that resembled the Midget Coupé, and Abingdon's resident model maker, Harry Herring, made up a neat model finished in red, its roof subtly raised, and a taller windscreen installed. Thankfully, this model survives in the British Motor Museum collection at Gaydon, even if its subsequent treatment at Abingdon seems to have been less than reverential. Enever took the model to show George Harriman, who had ordered the Frua MGA about six years earlier, but was told to take the project to Pininfarina, BMC's favored stylist of the day.

In short order a Chelsea Gray MGB roadster was shipped off to Turin, along with Stimson's drawings, and the eventual outcome was a sharp-looking running prototype in steel that was, to all intents and purposes, the form of what would become the MGB GT, launched in September 1965. The changes that the design studio had wrought were subtle but truly effective. They had squared up the roof, giving it crisp, sharp creases and a more rectangular windscreen, as well as taller side-window glasses. At the rear, a large opening tailgate—long before anyone spoke of a "hatchback"—opened to reveal luggage space that was quite generous for a sports car.

Jim Stimson insisted to the author that "the height of the windscreen was increased at the very beginning by us and not by Pininfarina. The only significant alteration that Pininfarina made was to add the feature line—the sharp crease—around the roof, which I must admit improved the appearance immeasurably." Clearly the opening tailgate was another major change: The first Pininfarina sketches did not show such a feature, but Stimson was perennially frustrated when others claimed that Pininfarina deserved sole credit for the higher roofline.

Painted in the pale Metallic Green that Enever favored for MG prototypes, the finished car returned to the U.K. in the summer of

A half-step toward an all-weather MGB: The factory offered this optional hardtop, manufactured by Ferranti Laminations of Bangor, North Wales, and first seen in public at the 1963 12 Hours of Sebring. Many dealers in overseas markets ordered new MGBs with these hardtops fitted, realizing that they were a cheaper option for import. And even if the customer didn't want the hardtop, they'd have little difficulty selling it to someone else. *Plain English Archive*

CBW 55 B

OPPOSITE TOP: Around the same time MG was planning its own hardtop coupe version of the MGB, an enterprising Belgian gentleman named Jacques Coune began offering hand-finished conversions based on his own design. This car was specially commissioned by Walter Oldfield, managing director of the Nuffield Press, largely as subterfuge so that BMC could study the car for themselves. This remained the only right-hand-drive Coune MGB; today it resides with the Coune family. *Enever Family Archive*

BELOW: Sergio Pininfarina provided this photograph exclusively to the author. This is a fully finished MGB GT prototype in Pininfarina's Turin studio. The car was extremely well received by all who saw it at Abingdon and Longbridge, and survives today in a private collection. *Pininfarina SpA*

1964 and impressed everyone who saw it, from Harriman (by now BMC chairman) down. The process of translating the new shape to tooling was a rapid one, aided by the Mylar drawings furnished by Pininfarina; Pressed Steel was rewarded for its quick work by receiving the assignment of not only executing basic pressings and subassemblies but also performing the complete body-in-white work at Swindon. Painting and partial trimming took place at Cowley, with final assembly then being completed at Abingdon alongside the painted and trimmed MGB roadster bodies, which for the time being still came from Coventry.

The MGB GT was undoubtedly a success, its neat styling winning many converts, although perhaps inevitably sales were better in some markets than others; the United States still showed a strong preference for open sports cars in this price range. Thornley and Enever were justifiably proud in any event, the former proclaiming that, in his view, the new car was one in which any company managing director would be happy to arrive at work. In the meantime, changes to the whole MGB range brought in push-button door handles and a slightly more refined five-bearing version of the B-Series engine, an improvement shared with the new Austin ADO17 1800 saloon range. Exclusive to the GT, however, was a beefier Salisbury rear axle.

Production of the new GT was ramped up, and in early 1966 a statement was issued to the

ABOVE: Syd Enever and Jim Stimson's scale model proposal for an MGB Coupé predates the definitive Pininfarina prototype. Contrary to popular belief, the windscreen height and roofline were both raised, but there's no doubt that Pininfarina's deft styling touches enhanced the end result immensely. The model is held at the British Motor Museum in Gaydon. *Enever Family Archive*

The MGB GT became the MGB/GT when launched in the United States for 1966; the idea came from BMC U.S. sales executive Bob Burden. Publicity for this car was overseen by Marce Mayhew, who decided that placing attractive female models on the roof of the car helped underscore how different it was from the usual open sports car that was more familiar to the U.S. audience. *Marce Mayhew*

effect that even John Thornley himself could not lay his hands on an MGB GT because they were all going for export. The preference in the United States remained the open roadster, though: People who wanted a closed-roof two-seater could buy a new Ford Mustang for roughly the same cost. To bolster sales in the United States, a "limited edition" of what was being marketed locally as the MGB/GT was created by the local sales team in 1966, offering a number of dealer-fitted options as standard with the package. It was an exercise that successfully moved surplus stock, and such cars are now prized, even if technically their creation and provenance had little to do with MG back at Abingdon.

MORE THAN MEETS THE EYE: THE MGC

Syd Enever had long harbored an ambition to recapture the appeal of a six-cylinder MG. Such a vehicle had last been seen in the MG SA and WA ranges, which died out with the coming of World War II. The 2.2-liter overhead-cam Wolseley 6/80 engine that was adopted briefly in destroked two-liter guise for Goldie Gardner's last record-breaking forays also had some appeal, but with the BMC merger that engine was destined to be culled. More promising was the new 2,639 cc Morris Engines large C-Series six, which appeared in the Morris Isis and Austin Westminster, and then the Gerald Palmer–designed Wolseley 6/90, which replaced the old 6/80. Enever had an MG Z Magnette fitted with one of these engines and used it quite extensively, including for family outings.

Then came the evolution of the big Healey, with the 100/6 being a slightly stretched 100/4 with the new Morris Engines six up front in place of the venerable 2,660 cc Austin four (itself an offshoot of the old Austin D-Series Six truck engine). By the time Austin-Healey production moved to Abingdon, with Enever as its guardian, the MG designer turned his thoughts to a future upmarket sports car that might conceivably replace the Healey while allowing a new MG six-cylinder. If that idea had been sanctioned, then perhaps greater success might have been realized. The high tooling costs of the MGB, the need to focus on the other sports cars, and other factors conspired to ensure that the MG six never got traction as a discreet project.

As John Thornley told the author, "During 1957, production of the Austin-Healey 3000 was transferred to Abingdon from Longbridge, and design responsibility fell on us. The immediate effect of this was that the future MGB became not only the basic successor to the MGA but to the Healey 3000 as well. It is this dual requirement which accounted for the large gap between the back of the radiator and the front of the engine in the early MGB. It also means that the car is

ABOVE: The MGB GT was regarded by Sergio Pininfarina as one of the best outcomes of his company's collaboration with BMC. *Martin Williamson*

LEFT: The MGB GT was exhibited at a number of major international motor shows using this novel display: The two halves of a sectioned car were moved apart and together on a large circular dais. This photo is from Earls Court, but the same stand was shipped to New York and a number of other European shows. The sectioned halves (but not the rest of the mechanism) are a major attraction today at the British Motor Museum. *Author archive*

ABOVE: The MGB GT featured this simple arrangement of badges on the tailgate—the latter in effect a sports hatchback, before the term had been invented. *Martin Williamson*

BELOW: The interior of this Tartan Red MGB GT Mark I, with red leather trim, is not all that different from that of a contemporary open MGB. *Martin Williamson*

some six inches [15.25 centimeters] longer than it might otherwise have been, a contribution to the impeccable handling of the B."

For a brief period, toward the end of the 1950s, there was also the prospect of an all-new and more compact range of V-4 and V-6 engines, of roughly two- and three-liter capacities, respectively. V-4 prototypes were tried in MGAs, and both the V-4 and V-6 were schemed into the ADO23 on a purely packaging basis—this partly explains the wide MGB engine bay. In the end, though, the new engine project was shelved in favor of revamping the A-, B-, and C-Series units.

What happened instead of creating a unique platform six-cylinder sports car was a plan, begun before the MGB was even launched, to stretch that new model by creating a six-cylinder version, with the thought that both MG and Austin-Healey variants could be created. Accordingly, the Healey concept became ADO51 and the MG was dubbed ADO52. Prototypes were built over the following years,

and, in addition to the C-Series, MG also looked at a BMC Australia six-cylinder version of the B-Series four, which had been designed in the U.K. but never used there. Various other ideas were explored. The challenge common to all involved maintaining the balance and purity of the MGB original while finding a way to accommodate the larger powertrain.

In the event, the engine selected was a revised version of the C-Series, which Morris Engines promised would be much lighter, and an MGB monocoque reworked so much under the skin forward of the passenger compartment that it was almost a new car beyond its styling. The much bulkier engine, with a choice of a new, all-synchromesh manual or automatic Borg-Warner gearbox, meant the front suspension was completely new, while the radiator had to be moved further forward. Squeezing the engine under the bonnet necessitated a broad bulge, which was crowned with a transverse trim strip and surmounted on top of the main square promontory with its own teardrop bulge to accommodate the carburetor arrangement on one side of the engine.

Intended as the MGC to stand with a largely identical Austin-Healey 3000 Mark IV, the latter was canceled at a late stage when Donald and Geoffrey Healey saw the prototype and put their foot down. Donald tried to get George Harriman to agree to a low-volume "soft tool" option to create different fenders, and to consider, once available following the 1966 BMC merger with Jaguar, the adoption of the Daimler V-8. But it was not to be. The MGC was launched in September 1967 with a slightly desperate strapline—"More than meets the eye"—minus its Healey equivalent.

The MGC arrived, and with it a much-improved MGB Mark II, just as BMC's fortunes and its public image reached a low ebb. The similarly powered ADO61 Austin Three-Litre appeared at the same time, and both seemed predestined to receive bad press. Add the fact that the non-Abingdon team responsible for the press launch used the wrong front tire pressures, which badly affected handling, and the poor MGC got off to a sticky start in the public's eye. When the U.K. press began to criticize the car's heavy handling, overseas journalists followed suit. Beneficial changes were made for the 1969 model year, but by then new masters were in control, the future of the MGC was in doubt, and the press garage failed to offer the much-improved cars for test.

In the meantime, almost unnoticed, the last MG Magnette Mark IV was built, bringing to an end the rear-wheel-drive inline engine MG saloon, at a time when the likes of the Ford Cortina/Corsair and Vauxhall Victor (including the sportier VX4/90) were showing how successful such a formula could be if handled with aplomb.

As an engineering exercise, the MGC was far better than was ever given credit. Syd Enever and Roy Brocklehurst made a good effort, which could and should have been better, had their pleas to BMC management been heeded. The Downton-tuned MGC, along with Daniel Richmond of Downton's personal Austin Three-Litre, showed just how potent it could have been.

Announcing the most important sale in MG history...
THE FIRST ANNIVERSARY
MGB/GT SPECIAL
Right now MG dealers across the country are commemorating the anniversary of the first popular-priced authentic GT—the MGB/GT.
In honor of the occasion, they're offering MGB/GT's in a special edition. Each one will have all the extras MGB/GT's come with plus a special anniversary package.
Only 1,000 First Anniversary MGB/GT Specials will be available in the entire U.S.A. So don't wait.
See your MG dealer before he runs out.
THIS SPECIAL ANNIVERSARY PACKAGE FREE . . .
Official plaque in full color designating the GT as one of only 1,000 Specials.
16-inch wood rimmed steering wheel with matching Australian coach wood shift knob.
Vibrationless non-glare racing type wing mirror.
. . . PLUS ALL THESE EXTRAS AT NO EXTRA COST ON EVERY MGB/GT:
Large electric tachometer and full sports car instrumentation.
4-speed gear box with short throw stick shift.
Oil cooler for better engine performance less engine wear.
Heavy-duty, competition-proved suspension for remarkable road-holding.
60-spoke wire wheels. Center lock type. Retail value over $100.
Twin S.U. carburetors for quick acceleration.
Hydraulic disc brakes with big 10¾-inch discs. Self-adjusting. Fade-free.
Fully adjustable English leather bucket seats.
MGB Convertible, MG Midget, and Austin Healey Sprite also available. See your MG/Austin Healey Dealer Now.

Sales of the MGB/GT in the sports-car-hungry U.S. always lagged behind those of the open-topped roadster, so BMC's Mike Dale hit on the idea of a limited edition: the MGB/GT Special. Most of the special aspects were added locally rather than at Abingdon. In any event, the project was deemed a success. *Author archive*

much more than meets the eye!!!
advance information
MGC
SPORTS & GT

By January 1969, though, the new management team were considering discontinuing the MGC while keeping the saloon, and by May it had been agreed that the model would be dropped, leaving a clear field for the new Triumph TR6 and the forthcoming Stag.

The tragedy was that a rather wonderful motorsports derivative, the MGC GTS, had been developed at Abingdon. With its lightweight alloy panel work, beautiful flared front and rear wheel arches to cover the wider racing wheels, and a promising alloy version of the MGC engine, it could have transformed the road car beyond all belief. The story of that car is covered in the next section.

MG RACING AND RALLYING IN THE 1960S

At this point it is worth returning briefly to the motorsports story of the MG marque during the 1960s. The MGA, as we have seen, had an excellent track record in nearly every theater of motorsports in which it competed. For

ABOVE: When the MGC was first revealed to the press, BMC used the tagline "Much more than meets the eye." This and the rest of the campaign sought to entice onlookers to look beyond the obvious resemblance to the humble MGB, while at the same time perhaps admitting that the new model was not as distinctive as it might have been. *Author archive*

RIGHT: The MGB GT appeared in Mark II guise in late 1967, at the same time as BMC launched a new model with a larger engine. This stylish couple and their Mineral Blue LHD MGB/GT Mark II were photographed by Canadian Marce Mayhew for a contemporary U.S. advertising brochure. Note the all-amber lenses for the combined turn signal and passing light arrangement. *Author archive*

much of its lifetime, it had been one of the key ambassadors for BMC's motorsports team.

By the start of the decade, however, there were other BMC products that had rapidly assumed prominence in the corporation's sporting activities. The rugged and potent big Healey had evolved to become a powerful giant of the rallying world, and meanwhile the Mini Cooper S had emerged as a remarkable, pocket-sized giant-killer in the same arena.

Suddenly, as these other models came to the fore, MG was no longer the de facto focus for the BMC Competitions (Comps) program. Nevertheless, as BMC's primary sports car marque, MGs were obviously a part of the effort. The MG Midget and MG1100 were quickly put to work: In the January 1962 Monte Carlo Rally, an MG Midget (registration number YRX747, rally number 44) was driven to thirty-third place by the rallying clergymen Rupert Jones and Phillip Morgan (the Morley twins had finished in twenty-eighth place in the previous Monte in the MGA Coupé 151ABL, bearing rally number 314). The MGB was not launched until much later in the year; wisely, nobody seems to have given the MG Magnette Marks III or IV any credence as serious motorsports prospects.

The following year, in the January 1963 Monte, an MG1100 was driven by Raymond Baxter and Ernie McMillen (registration number

ABOVE: Resplendent in Tartan Red with black leather seats and the de rigueur wire wheels (15 inches [38 centimeters], a size up from those of the MGB), the new MGC was a smooth motorcar with its new seven-bearing, three-liter engine. Unfortunately, it arrived at a troubled time for the British Motor Corporation and was unveiled to a moderately indifferent audience; a botched press launch didn't help matters. *James Mann*

BELOW: The new engine in the MGC was shared with the new ADO61 Austin three-Litre, another car that deserved better than the reception it got. It was similarly undermined by the attitudes of some commentators. Contrary to what some believe, this is not the old C-Series engine that had seen service at BMC since 1954, but a substantially reengineered unit. Another detail that didn't help its reputation: its use of a heavy cast-iron block. *James Mann*

ABOVE: One rather special customer for the new MGC GT was Prince Charles (the future HRH King Charles III), who took delivery of his Mineral Blue MGC GT from University Motors. He brought it to Cambridge University with him. Seen here being prepared at Abingdon, the car survives at Sandringham, somewhat incongruously reregistered with a different number. *Author archive*

LEFT: Alongside the MGC there was the inevitable MGC GT (or MGC/GT in some markets). In the opinion of many, the MGC GT—especially with the Borg Warner BW35 automatic gearbox—was the best iteration of the MGC in terms of a continental cruiser. *James Mann*

399CJB, rally number 268) finished in sixty-sixth place, and the MG Midget YRX747, now bearing rally number 158 was driven in this year to sixty-ninth place by Jones and Morgan. But the real excitement among BMC fans came from the Mini Coopers driven by the teams of Rauno Aaltonen and Tony Ambrose, and of Paddy Hopkirk and Jack Scott, and the Austin-Healey 3000 of Timo Mäkinen and Christabel Carlisle. There was a single privateer MGB and eight ADO16s (MGs and Morrises) in the same event, all of which retired.

Much was hoped for the MGB at Sebring in March 1963, but, as we saw earlier, that race ended ignominiously. There was still hope for that summer's Le Mans race, where MG managed to support a quasi-private entry to circumvent the BMC racing ban that was still in place at this time. The formal entry was made by Alan Stuart Hutcheson, a larger-than-life character who had been racing with BMC for a couple of years. His team name, "Ecurie Safety Fast," was a combination of the MG motto and the name of the contemporary in-house monthly magazine aimed at Austin-Healey and MG owners and enthusiasts. Hutcheson's codriver was the irrepressible Ulsterman Paddy Hopkirk, who was already earning an international reputation for his prowess in rallying Mini Coopers.

The MGB they drove was a Tartan Red roadster (registration number 7DBL, race number 31). It featured an aerodynamic nose that had been conceived by Syd Enever and detailed by Jim Stimson, the draughtsman of the future MGB GT, and a white Ferranti hardtop with a longitudinal red stripe along the center of its roof. Hutcheson started the race, but a lapse in concentration sent him off the road at the Mulsanne Straight, leaving the car stuck in sand for nearly an hour and a half while Hutcheson dug it out (Le Mans rules forbade any external assistance in such circumstances). Once freed, the MGB continued, with sand everywhere inside the cockpit (according to Paddy Hopkirk) but otherwise no ill effects from the misadventure. Between the two drivers, the car achieved a maximum of 132 miles per hour (212 kilometers per hour) down the Mulsanne Straight and ended the event with a highly commendable twelfth place overall.

The next month, Hutcheson was involved in a serious crash at Silverstone in 6DBL, which resulted in a trip to the hospital for him and significant damage to the car. The remains of the MGB survived and were later reconstituted, as was often done in those days. September saw 7DBL, with the same distinctive rounded nose, competing in the Tour de France (race number 155) with the driving shared by Andrew Hedges and John Sprinzel. They had been doing well in this arduous set of circuit hops when they crashed out. As Sprinzel told the author, "Andrew had

MGB's debut at Le Mans in June 1963 was more rewarding than its turn at Sebring, if hardly less dramatic. Alan Hutcheson shared the drive with Paddy Hopkirk, but coming off into the sand early in the race nearly jeopardized their race.
Author archive

TOP: Syd Enever was encouraged to create a handful of special, lightweight MG Midget coupes for racing purposes, using alloy body panels glued and riveted onto an ADO41/47 base structure. Two of the three cars built were loaned to MG dealer and racer Dick Jacobs, who had crashed his EX182 at Le Mans in 1955. The cars became known colloquially as "Jacobs Midgets." *Author archive*

ABOVE: The businesslike cockpit of the MGB as prepared for Don and Erle Morley to drive in the 1964 Monte Carlo Rally. In the actual rally, they finished an impressive seventeeth overall, but understandably most of the international media interest was on the first-place Mini Cooper S of Paddy Hopkirk and Henry Liddon. *Author archive*

a little nap: Unfortunately he was driving at the time, so we spent most of the night in the clinic."

By this time, the sheer dominance of the Mini Coopers and Austin-Healey 3000s meant that the factory focus on MG within Europe was limited. MGBs were entered at Le Mans and the Marathon de la Route for a while, but this was generally to seek a class win and some positive publicity. BMC often reserved a full-page ad in certain publications that would be used to shout about some success or other. If there was nothing to celebrate, the advertising space could still be used to cover some more mundane BMC road car.

For the January 1964 Monte Carlo Rally, the Morley twins codrove 7DBL, now without its rounded nose, bearing rally number 83. They achieved a fantastic first in their class and seventeenth place overall, which was almost as noteworthy as contemporary Mini Cooper S victories. The celebration was tempered by the sad news that one of the BMC mechanics, Doug Hamblin, had been killed in a road accident as he transported a spare radiator intended for the MGB.

Sebring 1964 saw the task of race preparation and team management handed to BMC's Northern California distributor Kjell Qvale, whose mechanic Joe Huffaker took three very early MGBs from stock, rebuilt them with lightweight panels that had been shipped over from England, and likewise rebuilt the engines that Abingdon supplied. The cars were finished in red, white, and blue, with contrasting stripes, and all three—piloted by an all-American team—made good headway, although the white car suffered an axle failure and was forced to retire.

MGBs returned to Le Mans in 1964 and 1965. On both occasions they appeared as official factory entries, fitted with the aerodynamic nose cone but on different cars: The entry for 1964 was BMO541B, and that of 1965 was DRX255C, both red with the usual white hardtops. Race numbers were 37 and 39, respectively, with Hopkirk and Hedges driving both years. In 1964 the MGB won the Motor Trophy as the highest-placed British entry. Notably, as at Sebring, the MGB was increasingly being outclassed by more specialized machinery. Although Syd Enever harbored an ambition to return with a new design, Abingdon's love affair with Le Mans officially ended in 1965.

Across the pond, however, MGBs continued to feature not only in factory form at Sebring, but in many Sports Car Club of America (SCCA) events across the United States. They regularly dominated their class, a remarkable situation that brought new drivers into the limelight and persisted even after the end of the MGB's production story. The arrival of the MGB/GT in U.S. showrooms was too late for its entry in the Sebring 1965, though two roadsters with hardtops appeared in that race: a GT Class DRX256C and a Prototype Class BMO541B (race numbers 48 and 49) that finished in the thirty-second and twenty-fifth places overall). None of the cars were entered at Sebring the following year, although two MGB roadsters with hardtops, a Prototype Class 8DBL and a GT Class HBL129D (race numbers 44 and 59), finished thirty-fourth and seventeenth.

In Europe at this time, MGBs in the hands of privateers made significant headway in such events as the continental one-thousand-kilometer races and the 1966 Ilford Films 500 at Brands Hatch, but here the focus was on factory team efforts. There were still two big MGB performances in 1966. The first, Sicily's Targa Florio in May, saw Timo Mäkinen and John Rhodes conduct GRX307D, a car soon to be known as "Old Faithful." Wearing race number 64, they finished first in their GT class, while Andrew Hedges and John Handley drove JBL419D (race number 66) to finish second

LEFT: Like the previous "Varitone" ZB Magnette, the Farina MG Magnette Marks III and IV were commonly offered with the option of a duo-tone paint scheme. This required an additional stage in painting the bodyshell prior to assembly. *Author archive*

FOLLOWING PAGES: Prior to the 1968 Marathon de la Route, both the factory MGC GTS cars—MBL546E and the newer one, RMO699F—were tested at the Thruxton race circuit. Although the driver's face is not clear in these photos, Paddy Hopkirk was one of those who took turns behind the wheel. *Stuart Spencer*

MBL 546E

in class behind their teammates. The other race of note, the Marathon de la Route rally, saw GRX307D again, now with Hedges and Hopkirk driving (and race number 47) to achieve an almost unbelievable overall victory.

"It was not surprising therefore that the MG pit went just a little bit wild when No. 47 came flying over the brow to take the chequered flag," Peter Browning wrote in the monthly *Safety Fast!* BMC Competitions mechanic Den Green told the author, "It was unbelievable. I really didn't think we would be likely to even finish. Of course in rallying we used to do all sorts of things to keep going that weren't exactly in the engineers' books!"

For Sebring 1967, a lightweight version of the MGB GT finally made an appearance. This was LBL591E, in Tartan Red like all the factory roadsters to date, wearing race number 30 and driven by Hopkirk and Hedges; they finished eleventh overall. Finishing twelfth overall was a roadster with hardtop, GRX307D (race number 48), driven by John Rhodes and Timo Mäkinen.

Soon after Sebring, a prototype lightweight GT, wearing registration number MBL546E, was seen at the 1967 Targa Florio. Boasting elegantly flared wheel arches, this new car was technically a prototype: It was in fact an early MGC GT, but with a two-liter B-Series four. It was hastily resprayed British Racing Green when the crew learned that this was a requirement for it to be allowed to race in its prototype class. Destined to assume its MGC guise later, when the production version was launched, MBL546E became known inside Comps as "Mabel." Wearing race number 230, Mabel completed nine laps but was not formally classified at the end due to the somewhat arcane Targa Florio rules.

Although roadsters remained part of the MGB racing story, the focus now shifted somewhat toward GT bodies, whether B or C based. The first GT, LBL591E, was badly damaged in shipment, necessitating a rebody into a new British Racing Green shell, after which it appeared at Sebring again and then at Targa Florio. As the original shell was later rescued from the skip and rebuilt, the LBL591E holds the peculiar position of being in essence two separate cars: One remains in the U.K., while the other resides in North America.

Although there were plans for several more, only two MGC GTS works cars were formally completed at Abingdon and raced at Sebring and Nürburgring, one of them later fitted with the alloy-block three-liter engine.

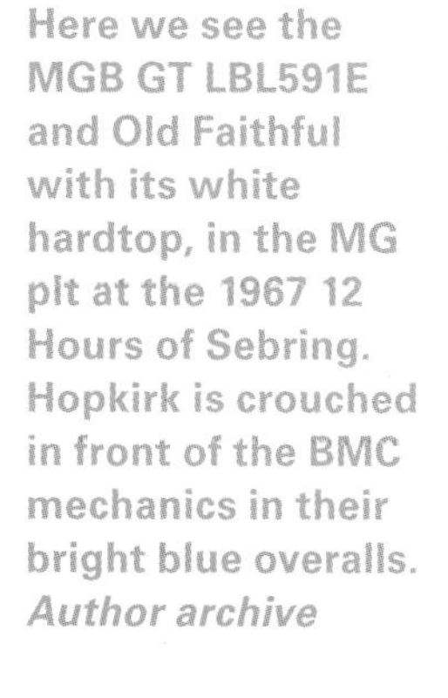

Here we see the MGB GT LBL591E and Old Faithful with its white hardtop, in the MG pit at the 1967 12 Hours of Sebring. Hopkirk is crouched in front of the BMC mechanics in their bright blue overalls. *Author archive*

MG's BMC masters hadn't openly considered replacing the MG Midget and MGB; it was just that funding was never prioritized. Instead, the urgent focus had become compliance with the raft of new U.S. federal regulations that threatened to undermine the sports cars already for sale. In fairness, these challenges also bedeviled MG's rivals at Triumph, part of the Leyland organization. This is the EX234, an attractive Pininfarina-styled prototype with removable hardtop and BMC's Hydrolastic all-independent suspension. It could have replaced both the Midget and the MGB, but this was not to be. *James Mann*

Despite promising performances, they could not save the MGC, and before long Abingdon's Competitions Department also succumbed to the corporate axe. Sebring 1968 saw Mabel in definitive MGC GTS guise (race number 44), racing alongside the second iteration of the lightweight MGB GT LBL591E (race number 66). Driven by Paddy Hopkirk and Andrew Hedges, the MGC GTS finished first in its class and placed an impressive tenth overall.

The next outing for the MGC GTS was October 1968's Marathon de la Route. Two cars took part, MBL546E and RMO699F, the latter fitted with an experimental alloy engine. The newer car (race number 5) overheated and was forced to retire, but, remarkably Mabel (race number 4, with drivers Julien Vernaeve, Tony Fall, and Andrew Hedges) lasted the eighty-four hours and finished sixth overall, first in its class.

The MGC GTS' successes mattered little with regard to corporate changes far removed from the competition. Within days of the race's conclusion, BMC's Sir George Harriman had resigned and the new management announced the closure of the Comps department—but not quite in time to kill the following spring's Sebring entry.

That race, in 1969, saw two MGC GTSs take part: the MBL546E (race number 36), accompanied by RMO699F (race number 35, with Hopkirk and Hedges in the newer car), while the MGB GT LBL591E also raced (race number 62). The GTS cars finished sixth and seventh in their class. It was an encouraging result, but, as the Competitions department had closed, any future MG entries for the next decade—and on the SCCA race circuits for considerably longer—were all thanks to the efforts of privateers.

BIGGER IS BETTER: THE LEYLAND MERGER

The prevailing view in the mid-1960s, especially that of the contemporary government, was that the consolidation of the motor industry through mergers and takeovers was the best way to secure its overall survival. Harold Wilson became Labour prime minister on October 15, 1964, then gained further political clout with a second election victory in 1966. Wilson was a fan of the five-year plan, beloved of Communist states, and was a supporter in principle of nationalizing large industries.

To encourage growth and the spread of industry, Wilson created the Industrial Reorganisation Committee and promoted technological growth. He was impressed by the dynamic chairman of the Leyland Motor Corporation, Donald Stokes, and frustrated by the fractured industrial relations at the much larger British Motor Corporation, of which MG was of course a small part. In 1966, a damaging labor dispute at BMC reinforced the problem as Wilson saw it. It was hardly a coincidence, then, that BMC and Leyland began the long series of talks and, sometimes, talks about talks, from

Go turismo.
Go where the camions smoke and rumble down the Routes Nationales. Head where the scented sun oils glisten and the cicadas creak under the cypress trees. Promote yourself to the glassbacked, fastbacked, sleek-lined grand touring world of MG.
Go MG–for the sport of it.
MG MIDGET
MG 1300
MGB
MGB/GT
MGC
MGC/GT
MG

An MGB GT advertisement from January 1969. The cars looked hardly different from the ones built before the merger, but changes were well into the planning stages by the time this ad appeared. "Go MG for the sport of it" was the tagline. Despite being a U.K.-market campaign, the names MGB/GT and MGC/GT, normally reserved for the North American market, are used here. *Clive Richardson*

around the time of the 1964 Motor Show, just days after Wilson came to power.

The government's enthusiasm for consolidation was also championed by Anthony Wedgewood-Benn, who took control of the Ministry of Technology (popularly known as "MinTec") and worked behind the scenes to broker mergers. By this time, BMC's success with the Mini and 1100 was faltering, and it had also been experiencing calamitous losses at its Bathgate commercial vehicle plant in Scotland, the creation of which had been another of those political ideas to spread employment to remote areas. In the face of these challenges, the company had recently expanded to form British Motor Holdings (BMH) through the acquisition of Pressed Steel in 1965 and a merger with Jaguar and Guy Motors of Wolverhampton in the summer of 1966. The corporation, now the U.K.'s largest carmaker, seemed to have become a giant with feet of clay, and there was growing concern that it was at risk of collapse.

The leaders of Leyland (Stokes) and BMH (Harriman) were quite different personalities, but when the government tried to act as matchmaker, they both felt encouraged to look at combining their operations in some way. At the time when these corporate dances began, it was assumed by nearly everyone that BMC would have been the dominant partner in any relationship, but the ensuing negotiations were not helped by Harriman's declining health. (As we saw above, he would resign his chairmanship in October 1968, dying just a few years later on May 29, 1973, aged 65.)

BMH had focused on the supremacy of technology, ideas, and innovation, even at the expense of ultimate profitability. Dominated by people who felt that car making should essentially be fun, the business itself could not support that vision. "Fun" was not the number one priority of the highly organized Ford Motor Company, a business that still created desirable motorcars but also kept a close eye on the bottom line. Car making was complex throughout its various operations and, in part to address this complexity, the Austin and Nuffield wings of BMC's car business gradually became more closely integrated in their production. There were still separate management boards and sales organizations, the latter supporting different dealer networks depending on any of a multitude of brands that were on offer.

In the BMC sports car world of which Abingdon was a part, for the U.K. and some other markets, Austin-Healey sports cars were marketed and sold mainly through the Austin dealer network, while MGs went via the Nuffield chain. The alliances made by BMC with Jaguar and Pressed Steel could have bolstered MG, but it was a dream for the future at a time when much larger commercial challenges gripped the minds of senior management. There is insufficient space here to set out all the details of what happened, or exactly why—nor how circumstances could easily have led to a different outcome when both the key parties began to get cold feet. Suffice it to say that, when matters came to a head in January 1968,

BMH and Leyland announced that they were merging. Before long the union was confirmed and the British Leyland Motor Corporation (BLMC) came into existence in May.

For many in both organizations, the shock of this grand merger was hardly any less than what had been felt with the merger sixteen years earlier, when BMC was born from Austin and Nuffield. MG's champions were already frustrated by their parent company's apparent poor investments and were chastened by the rapidly growing effects of U.S. government legislation on sports car exports to the United States. Now they naturally worried that MG had joined the same parent company that owned their deadly rivals at Triumph. That company had been a critical part of Leyland's car-making side, and thus it had benefited from closer attention on planning and investment from its board. Indeed, Standard-Triumph had grown robustly since its acquisition by Leyland in 1961.

A wide range of Triumph projects were in production across a number of factories, the cars including nearly brand-new models such as the Herald, Vitesse, 1300, and highly regarded 2000 and 2.5PI, not to mention the Spitfire, GT6, and TR sports cars. The company was also engaged in a project to build and sell new in-house-design slant four engines (with the promise of a V-8 engine offshoot). They were working on tentative plans for an integrated automotive product range that offered a variety of sporting saloons and state-of-the-art sports cars.

MG, on the other hand, managed two successful but aging small to midsize sports car ranges that included a larger-engined sports car, the MGC, which, as we have seen, was foundering in the market; a fairly popular saloon, the 1100; and an antediluvian one, the Magnette. Nothing solid was in the pipeline beyond some face-lift proposals developed by BMC's new head stylist, Roy Haynes, who came over to the corporation barely three months before it fell into the arms of Donald Stokes.

Having presided over the Leyland rescue of the business in 1961, Stokes understandably had a soft spot for Triumph, an affection he seldom concealed. So it seemed likely from the outset that, if he could, he would push the Coventry marque's agenda over the Abingdon one. In an interview with the author in 1990, it was clear that he still believed some of the falsehoods that had been peddled about the relative sales success of MG and Triumph in key export markets, even years after the fact. It would not be long before the implications of these biases came to the fore, although, as we shall see, MG sometimes had unlikely champions in their corner.

A January 1969 photo of a U.S. -specification MG Midget. Note the left-hand steering and all-red taillamp lenses. *Marce Mayhew*

MGB/GT

6

Success Against the Odds: The Seventies

MG ENTERS THE BRONZE AGE

The 1970s began as a peculiar decade, coming after the bright and vivid 1960s, which had seen such cultural change and a sense of bright if often naive optimism. Fashions in the car world were not exempt from the wider changes in taste; color schemes slipped toward a new generation of what might be thought of as impure tones, often muddier hues, with a preponderance of browns, beiges, acid or bronzy yellows, orangey reds and murky greens, and the occasional purples.

In the MG world, these new color schemes coincided with an attempt in design to perk up the style of the MG Midget and MGB by dispensing with the classic chrome slatted grilles and replacing them with matte black recessed grilles with rectangular silver rings. This change made them vaguely resemble the recessed black grille of the 1968 second-generation Ford Mustang, popularly remembered for its starring role in that year's popular car chase film, *Bullitt.*

These MG changes were developed under the watch of designer Roy Haynes, who had been recruited to BMC in October 1967 after having delivered, at Ford of Great Britain, the neatly styled Mark II Ford Cortina. It was a car whose 1600E version had risen to become the fantasy of many of the U.K.'s young executives. Haynes was now

OPPOSITE: The 1970 MGB and MGB/GT featured a new recessed satin black grille modeled very loosely on that of the 1968 Ford Mustang and contemporary European sports coupe thinking. Gone is the traditional MG grille shape. The setting for this advertising shoot was the Old Rhinebeck Aerodrome in Red Hook, New York, home to a collection of World War I aircraft. *Marce Mayhew*

RIGHT: The route to the definitive "British Leyland MGB face-lift" was in stages. This photograph from January 1969 was taken at a time when the spruce-up might have been applied to both the MGB and the MGC. Note the B-Type legend on the sill tape, low-mounted BL badge, and matching satin black bonnet and grille. *Author archive*

BELOW: These beautifully detailed scale models depict a red roadster, the original Michelotti-styled Triumph Bullet, and the green coupe, the Triumph Lynx (which actually came first). The idea was to use the forthcoming Triumph slant four engine that was familiar to TR7 owners as well as an improved version of the Triumph six-cylinder. These were compelling offerings in the competition to find a new corporate sports car for British Leyland. *Author archive*

to deliver British Leyland's great hope in fighting back against the Cortina, under the project code ADO28. The resulting Morris Marina emerged in 1972 (in some markets, like North America, it was badged by Austin). By this time, however, Haynes had fallen out with the new BLMC management and left before the launch of the new saloon range.

Although the MG Midget, MGB, and MGB GT received their face-lifts in late 1969 for the 1970 model year, the MGC GT was not so fortunate, as the last example left the Abingdon production lines in September 1969. The car's end may have taken some by surprise: Documents seen by the author show that the MGC was being readied for the same changes, with five-spoke Rostyle wheels and other trim changes to echo those applied to the smaller sports cars. The newly formed Austin Morris Product Planning Committee had other ideas. Closely tied to the corporate team now ensconced in Marylebone, London, the group decided to kill off the six-cylinder MG. Only one component survived the cut, if temporarily, as the large Austin that had shared the same powertrain was kept in production for some time afterward.

At the turn of the decade, MG sports cars actively in production were approaching their ten-year anniversary. By industry standards they would have been replaced years earlier. Even the MGA had lasted only seven years and was considered somewhat aged and old-fashioned by the time it reached its end. It was not that Syd Enever and his team were short of ideas—far from it, as can be seen in their work on Hydrolastic sports cars and front-wheel-drive Midgets. Rather, it was a combination of insufficient funds at a time when BMC was working toward its merger and the need to keep the existing cars going in the crucial U.S. export market.

On December 31, 1969, Austin Morris and Triumph personnel were invited to a meeting, along with British ex-patriate Graham Whitehead and his U.S. sales team based in Leonia, New Jersey. On the agenda was a discussion of the North American market, which was key to the sports cars' future. The session included a reference to a live Austin Morris project, the ADO68 Condor, intended to be an MG coupe that was closely related to the

forthcoming ADO28 (Marina) saloon range. In some ways, Condor may have been too close to Triumph's separate Lynx project for comfort.

Both Triumph and MG factions were present at that December meeting. For Triumph, "Mr. King said that a Triumph Sports Car could be developed at low investment by cutting down a Triumph Lynx chassis and using Manx [future Dolomite] mechanical units. It was envisaged that the Lynx would replace the TR6 and GT6 and the 'Lynxette' the Spitfire. Mr. Swindle said that the preliminary costs of the Lynxette looked promising, and the model could take a slant 4 or V-8 if required."

In fact, the idea of Lynxette mentioned in these notes had been discussed at Triumph's own board meeting of October 10, 1969. By January 1970, what had begun as Lynxette had combined with Triumph's initial ideas under their Bullet sports car proposal, although this should not be confused with the later project of that name. For now the car existed only in model form; a full-size Michelotti-built running prototype of the original Triumph Bullet was not recorded in Triumph's experimental register as X817 until May 5, 1971, by which time Triumph had already built at least six of the related Lynx coupes. At the time, Triumph's main focus was on the development of their forthcoming small saloons (the Manx and Swift), in preparation for the Stag and, just like their rivals at Abingdon, dealing with the challenges of federal legislation.

Back to that December 1969 Austin Morris meeting. Over on the MG side, the Minutes

BELOW: One of the MG projects that did not survive was the ADO 68 Condor coupe, which would have been to the ADO 28 Marina what the European Ford Capri was to the Cortina (note the new Capri 1600 in the studio for this February 1970 review). There were several iterations of ADO 68. This one, with a small MG badge in the center of the grille, began as a Michelotti-built prototype that the Austin-Morris designers have clearly been reworking on the side nearest the camera to explore different ideas. *Author archive*

BOTTOM: The idea to make an MG out of a Mini never quite went away, although there remained doubts about how well the traditional sports car market would take to a front-wheel-drive car of this type. This is the ADO70 Calypso, designed in Longbridge but built in Turin by Michelotti. Its fate eventually became entwined with wider struggles within the parent company, not least the plans for a replacement for the base Mini range. *Author archive*

record that "there was some discussion on the need for a direct replacement for the MGB if the ADO21 (mid-engined Midget replacement) and the ADO68 (Condor) were available. Without having more detail of these programs it was not possible to reach a firm conclusion. Various individuals felt that the residual volume would not justify a replacement for the MGB; others felt that the traditional MGB market would be lost if we did not have a traditional open sports car." The ADO21 was a concept for a mid-engined sports car, originally conceived as a potential replacement for both the Midget and the Spitfire, but engineered and designed by Austin Morris as part of an intended program of collaboration across divisions.

Leonia's Graham Whitehead suggested his belief that, with an interior face-lift and more power, the MGB would continue to be viable for two to three years. The former Triumph man, Harry Webster, who was now in charge of engineering at Austin Morris, offered to investigate fitting the MGB with the new E-6 engine—an exercise that, history shows, got no further than another idea: to fit the MGB with the Triumph slant four engine.

Thus, at the end of 1969 and into 1970, the MGB seemed destined to continue as it was for the time being, based on the still-recent Haynes face-lift for the 1970 model year.

TURNING POINT

For a number of reasons, 1971 was a crucial year in terms of MG history. BLMC was emerging from a period of financial cutbacks, such as "Operation Survival," which was announced in late 1970. This plan meant that many of the corporation's bolder plans had been sacrificed on the altar of financial viability.

RIGHT: Another attempt at a new MG coupe in the Condor program. Although not visible in this shot, the studio clay even wore a Magna name badge on the rear—a nod to prewar MG history. In the event, a lack of funds and management prevarication led to this promising project being abandoned. *Author archive*

OPPOSITE TOP: The last new sports car project with any significant input from Syd Enever was this mid-engined concept, ADO 21, which would have used the Austin Maxi E-Series engine mounted amidships and driving the rear wheels. Although it was quite well regarded, the projected costs put it out of contention as a replacement for the MG Midget and Triumph Spitfire. It was abandoned in December 1970. *Author archive*

BELOW: This contemporary showroom poster for a U.K.-market MG Midget is of interest for several reasons. First, it exhibits the short-lived round rear-wheel-arch profile. It also shows the revised four-spoke Rostyle wheels introduced for the Midget at the same time. The color too is of interest: According to Austin-Morris designer John Stark, the Bronze Yellow was inspired by the 1969 Giallo Ocra (yellow ocher) popularly seen on the 1969 Alfa Romeo GTV. *James Mann*

MG, Triumph, and Jaguar sports car sales were all heavily reliant upon U.S. sales, but federal legislation there meant that all three marques had to cater to emissions rules, which sapped power and performance. Rules related to safety brought such elements as padded interior surfaces, seatbelt warning devices, and the ominous possibility of compulsory airbags and rollover standards. Taken together, these potential requirements threatened to completely wipe out the traditional open-topped motorcar. At the same time, there remained the unresolved matter of overlap between the MG and Triumph sports car ranges, a situation that was locked in place, mostly due to legacy concerns, within different sections of the British Leyland structure.

Separate schools of thought regarding sports cars below the Jaguar level had been in play for a while. On the Triumph side, as we have seen, the philosophy chiefly comprised an evolution of the Lynx concept, which had been in development for some time. On the Austin Morris side, two separate pieces of work were under consideration. One, a small Mini-based sports coupe called Calypso (ADO70), had been codeveloped by Longbridge and Michelotti, while

the other, a mid-engined sports car (ADO21), that had been conceived and styled at Longbridge as a replacement for the Midget/Spitfire. The latter was largely engineered at Abingdon to use the new E-Series engine. For this, MG chopped about an MGB GT to try out the running gear.

These various exercises were finally brought together in 1970 and assessed by various internal BLMC committees, including a Sports Car Policy Committee that met regularly and loosely connected Austin Morris (for MG) and Triumph. Despite his role as MG's chief designer, Syd Enever was no longer directly involved in the decision-making process. That function had been assumed on the MG side by Charles Griffin.

On November 16, 1970, the Sports Car Policy Committee met to view prototypes and full-scale models of the Calypso, ADO 21, and the Triumph Lynx, together with a quarter-scale model of Bullet. The minutes of that meeting recorded that "drawings and layouts of all models were seen and proposed specifications compared." It should be reiterated at this point that these early iterations of Bullet and Lynx were completely unrelated to the later projects that shared those names.

As far as each model was concerned, "all members agreed that Lynx was very attractive and acceptable and that it would absorb most of the TR 6 volume and compete with the Opel GT, the VW/Porsche 914/4 and Datsun 240Z. . . . however, it was felt that it was too expensive to be a replacement for the MGB which is currently priced $500 below TR6." As for Bullet, "it was felt that Bullet was too expensive and undistinguished in style and concept to be a serious contender as a replacement for Spridget and Spitfire in the cheap fun car market." Likewise, "ADO 21 with the 2.3-litre 6-cylinder engine, was also too large and expensive."

As we have seen, ADO21 had been originally intended for MG to function as a Midget, but the Abingdon engineers cunningly scoped the possibility of including the larger E-6 engine, its supporters hoping that the concept might bridge the Midget and MGB sectors just as had been planned for the earlier EX234 Hydrolastic sports car. In so doing, it could perhaps be argued that they had simply helped undermine the economic argument.

At that November 1970 meeting, it was concluded that "both models would be examined by their respective designers with a view to coming closer to market requirements for a two-seat sports car to sell at under $3,000 (in present MGB sector)." Both the Austin Morris exercises had started out as ideas to replace the ADO47 Midget. Unfortunately, the Calypso became

To the untutored eye, this may look like a dull-finished Triumph TR7. Look closer and you can spot the MG badges that were added as showroom jewelry to dress up the clay model. Lord Stokes was impressed by the design when he saw it at a viewing in July 1971. He decreed that it should be the basis, not of a new MG but of the TR7, code-named "Bullet." *Author archive*

complicated by debates over the future of the Mini on which it was based, along with lingering doubts over whether it truly offered a credible sports car replacement for the Midget. As noted by the Sports Car Committee, when a cost analysis was run, the mid-engined ADO21 proved too expensive for the intended $2,000 Midget market.

There had been no recent work on a specific replacement for the MGB. This was largely because the U.S. marketing team, led by Graham Whitehead, had stated that the Midget was the car that needed replacing first. Clearly the conclusions at that November meeting revealed a brewing standoff between a revised ADO21 and an enhanced Bullet. An almost throwaway comment at the November meeting offers some sense of a fallback option: "A third alternative discussed briefly was to face-lift the MGB using the Triumph slant-four engine with four-valve head." Needless to say, that idea fell on deaf ears.

ABOVE: George Turnbull was the director in charge of British Leyland's Austin Morris division, hence his role overseeing MG affairs. Here he appears at Abingdon celebrating the 250,000th MGB, a Blaze (orange) U.S.-specification MGB/GT, destined to be given away in a prize competition. The car survives. *Author archive*

LEFT: Just a couple of weeks before Turnbull turned up to celebrate the landmark quarter-millionth MGB, the man who deserved the lion's share of the credit for the MGB retired. Syd Enever's last day at Abingdon was at the end of April 1971. *Author archive*

It's sometimes forgotten that the MG Midget and MGB were assembled in a small number of overseas markets, using parts kits from the U.K. in addition to some local components to defray costs and comply with import tariffs. Australian assembly of the MGB ended on November 6, 1972, and the occasion was marked by a bizarre mock funeral. Leyland Australia did not survive much longer. *Author archive*

In the meantime, a fact-finding mission was sent to the United States in late 1970. Led by Spen King, essentially a Rover and Triumph man, the group learned what the local market appeared to need: It was not a mid-engined sports car design. This was concerning, as both MG and Rover had been working on something along those lines: the ADO21 and P6BS, respectively. Instead, the market analysis revealed that a more traditional layout, like the new Datsun 240Z and the Michelotti-designed Triumph Bullet/Lynx project, would be attractive to the American buying public. It should offer a more modern style that fit the times, which meant Bullet was not the model they desired. Contrary to some subsequently edited versions of history, Austin Morris chairman George Turnbull reported that, of the two sports car makes, MG was the one that needed to be retained at all costs.

These facts were laid out at a meeting of the Sports Car Policy Committee on January 5, 1971, which was attended by both Austin Morris and Triumph personnel; tellingly, no one from Abingdon had been invited. Spen King reported that "the forthcoming safety regulations are so severe that it does not appear possible or [economically sound] to engineer more than one model to meet them," adding that "on any new model, the regulations will have a major fundamental effect on styling." He went on to say that "the MG marque should be retained even at the expense of Triumph if it becomes necessary to limit products to one model." He made further detailed points, but in conclusion offered his view that "the way to go was with a straightforward, conventional (FE/RWD) car with removable roof panels." The comment about MG being effectively more important to the sports car debate than Triumph is a particularly striking one. Needless to say, this message was ignored and subsequently forgotten as vested interests took hold.

At the same meeting, we learn that "styling sketches were viewed in the Studio showing about 20 alternative themes. Nos A19, A20, A23 and A26, based on Triumph Bullet mechanicals, were selected for further development. The plan was for one bodyshell and chassis to be offered with the Triumph slant four, 1850 cc two-valve engine as MG, and with the 2000 cc four-valve as Triumph. Styling was asked to work up full-scale line

Not a great deal changed for 1972, other than new colors and minor trim changes. Advertising director Marc Mayhew set this contemporary photo shoot at the Conservatory of Flowers in Golden Gate Park, San Francisco. *Marce Mayhew*

TOP: Just the thing to park up alongside your private yacht: a Harvest Gold 1972 MGB/GT. *Marce Mayhew*

ABOVE: U.S. sales and marketing for the MG range were superb in the 1970s. The TV ad that saw an MGB parachuted to the Californian desert and then driven off ahead of its rivals won a major national award. The story behind the ad was no less exhilarating: On the first shoot, the MGB's parachute failed to open and the car crashed to the desert floor. Thankfully the team had a second car in reserve, hence the annotation on this photo by Marce Mayhew: "Take Two." *Marce Mayhew*

and tape drawings." So, in other words, a new "corporate sports car" was to be developed using Triumph running gear for both marques.

In the meantime, closer to the day-to-day world of production and sales, the MGB was about to notch up an impressive statistic: The quarter-millionth example went down the line in May 1971. It was an occasion marked with a visit from Austin Morris chairman Turnbull, who posed alongside the actual car, a U.S.-specification MGB GT, which was destined to be given away in a competition being organized in the United States. MG sales were still strong, and the point made by Graham Whitehead that the MGB still had a few years left appeared to be valid, even if the motoring press was beginning to suggest that a replacement was overdue.

One person who might have once expected an invitation to such an event was missing. Syd Enever had retired on the last day of April, following his sixty-fifth birthday. His old friend and mentor John Thornley had already retired two years beforehand at age sixty.

If it had not already been obvious that MG's future was being determined from afar, from the summer of 1971 on it became abundantly clear. Even so, while plans were being developed in secret, MG sales in North America remained strong, reinforced by the fact that there was a shrinking choice of open sports car on the market. A popular marketing strapline in the United States described MG as "The Sports Car America Loved First." It was a simple, affectionate statement of what was in essence a simple truth. Shamefully, some of the people in charge of British Leyland remained frustratingly myopic about that message.

LOADING THE BULLET

The debate about MG and Triumph and their future direction as British Leyland's small and medium sports car lines came to a head in the summer of 1971, with Lord Stokes himself sitting in on the final deliberations. With ADO21 having been rejected, and there being no conventional rear-wheel-drive Austin Morris ready beyond the projected Condor coupe, Triumph was clearly ahead in the face-off.

Hoping to stay in the styling race, Canley reworked its earlier Bullet design, giving it pop-up headlamps and bolder bumpers, which amounted to a look curiously close to that of the mid-engined VW-Porsche 914, which had made its world debut at the 1969 Frankfurt Motor Show. Triumph's modern slant-four engine, already developed for use

ABOVE: The privately built MGB V-8 conversions by Kent-based former racer and entrepreneur Ken Costello drew the attention of MG and BLMC HQ. Before long, Costello was commissioned to convert a car for Lord Stokes. This photograph is of the engine bay of that actual car, a Harvest Gold MGB GT. *Author archive*

LEFT: The 1973 Earls Court Motor Show in London saw the public debut of the new MGB GT V-8, displayed in a novel way with a sectioned engine beneath the elevated car on the stand. Despite plans to do so, the V-8 MG was never exported, and sales were hampered with the outbreak of war in the Middle East. *Author archive*

For the genuine factory MGB GT V-8 engine installation, MG's long-serving development genius, Alec Hounslow, devised a unique transverse twin-SU carburetor setup that allowed the engine to fit without the need for a bonnet hump like that of the MGC and the Costello cars. *Author archive*

by an external customer, SAAB, and destined for use in future Triumph saloon models, met anticipated U.S. federal emission levels with greater ease than the aging B-Series powertrain. Bullet in this form may only have really suffered from outdated packaging and doubts about its ability to meet future federal safety standards.

Longbridge, now the de facto center of MG design in addition to its principal Austin and Morris brief, had recently finalized the style of the ADO28 Marina (launched on April 27, 1971) and was hard at work on the much-anticipated new Austin (ADO67), which was to be a successor to the ADO16 "1100," alongside a replacement for the ADO17 Austin, Morris and Wolseley 1800 & 2200 range... (which would become ADO71), among other projects. It would have been a surprise if, having designed both the abortive ADO21 and ADO70 sports car concepts, Harris Mann and his colleagues had not been invited to participate in the submission of ideas for a new corporate sports car until the January meeting described above.

The Austin Morris design studio team created a striking wedge-shaped sports car model in clay, which looked like it was mid-engined but included a conventional sports car powertrain: It fit almost perfectly the idea described by Spen King at that earlier meeting in January. Dubbed the Magna and wearing MG badges—"styling studio jewelry," in Mann's words—the bold concept took Lord Stokes's eye when he finally saw it on July 9, 1971. He decided on the spot that this was to be the shape he wanted. The only catch was that it would be primarily a Triumph, the TR7, rather than an MG, with a possible allowance that the latter might follow in due course.

It is fair to say that most of the Canley team were horrified. They had felt confident that Stokes, their protector, would have gone for their updated Bullet proposal, or at least chosen a different clay model that had been prepared as an insurance policy by their design consultant William Towns, but Stokes flew away in his helicopter having made the decision. From MG's perspective, all their worst fears seemed to be coming true. For British Leyland's corporate sports car, the future seemed to be wedge-shaped with a Triumph label.

One witness to this was Nick Carver, who told the author: "Lord Stokes was there when the decision was made: There was a consensus that the Triumph style was too conventional. We needed something less dull and traditional. But in retrospect, maybe that wouldn't have been such a bad thing—and some others did feel that the TR7 [Harris Mann style] was too extreme. In the event it was preferred because it was 'different.' These sorts of decisions are rarely as strongly reasoned as you might like them to be—there is always a compromise consensus."

Triumph's compensation was that their engineering package was considered the only game in town. The next-generation Bullet would be engineered by and powered by Triumph, and almost certainly built mostly in a Triumph factory. In an unusual wrinkle, George Turnbull himself had made the case for an MG version of Bullet that would be built at Abingdon. Stokes gave the idea short shrift, and by March 1972, the die had been cast. Stokes decreed that all future Bullet sports cars, whether badged as MG or Triumph, were to be built at Speke, Merseyside. Back at Abingdon, something else had already arisen to keep the in-house development team excited: They had been asked to fit the Rover V-8 into the MGB.

MG MEETS ROVER: THE ALLOY V-8–ENGINED MGB GT V-8

Even before the MGC had reached production, discussions took place about changing the three-liter, cast-iron straight six for something else. Syd Enever had favored a shorter-stroke,

2.5-liter version with an alloy block and head, and there were some tentative thoughts of using the Daimler 2.5- or 4.5-liter V-8 engines, which Jaguar brought to the party when it joined forces with BMC. What many enthusiasts were even more excited about was the lovely alloy 3.5-liter V-8 that the Rover Car Company had acquired from General Motor. This had appeared in two 1967 models, shortly before the British Leyland merger suddenly placed MG and Rover in the same stable.

As early as December 1968, there were tentative plans to squeeze a V-8 into other vehicles. In the next few months, the Triumph Stag and the Rover 3500 units were both put forward as candidates for use in other models. The V-8 was also mentioned in the same breath as the ADO28 (Marina) and ADO68 (Condor) program, while work was undertaken to shoehorn the Rover V-8 into the large Austin ADO61 saloon. Indeed, Harry Webster drove a running prototype of the latter and wanted to consider it as a development of the three-liter.

Kent-based racer and car converter Ken Costello was not the first person to successfully squeeze the Buick (or similar Oldsmobile and Pontiac) V-8 into an MGB engine compartment, but he was nevertheless the first to make a business out of such conversions. In the process, he drew considerable attention to himself, inviting some of the leading U.K. newspaper

ABOVE: As part of the 1972 design studies that were looking at both impact and potential rollover legislation, MG worked through a range of options, including this idea of a rigid bumper on telescopic shock absorbers. In the end, the polyurethane "rubber bumpers" won out. *Jim O'Neill*

BELOW: Just as the MGB saw the adoption of large black rubber bumpers, irrespective of the sales market, so did the low-volume MGB GT V-8, sold only in the U.K. Here is a new 1975 MGB GT V-8 at the Earls Court Motor Show. By the time of the 1976 show, the MG V-8 was no more: British Leyland was actively promoting the new TR7. *Author archive*

motoring correspondents to try out his "Costello V-8" for themselves. Business took off in 1970 and 1971, and before long the instruction came down from BLMC HQ to look into brokering an arrangement of some kind. Costello received a letter inviting him to bring one of his cars to Longbridge: Being the bullish individual he was, he simply drove there without an appointment and called the staff out of their offices. Evidently the management were impressed and an order followed for Costello to build a car for them. In due course, Lord Stokes reviewed the car and discussed options, while he covertly gave instructions for MG to build their own MGB V-8.

By September 1971, the MG design and development team were progressing well with their own studies, and there was some degree of confidence at the Austin Morris product-planning level that the project had a strong chance of going forward, with certain provisos. Initially at the planning meetings, it was implied that tentative dialogue with Costello aimed to allow him limited support, but this switched to BL deciding to build its own car instead, and strangle support for the outsider. This meant that management were impressed with the prototype built under Terry Mitchell's supervision, albeit the design still had a few teething issues. Webster told the committee of "the completion of the first car, which in performance, road holding and noise insulation looked very promising. It was planned that three further cars should be built to right-hand and left-hand drive specifications to cover pollution and safety areas. The first car had been through a quick check on pollution requirements and was found to be completely unacceptable. All design and development is being carried out at Abingdon."

As we saw earlier, only four months earlier, Turnbull had smiled for the cameras next to a landmark 250,000th MGB GT as it rolled off the line at Abingdon. Now he duly gave permission for engineering to go ahead, saying that they should "continue with the project providing that there was no serious infiltration into resources which may be required on other projects and that expenditure was kept to a minimum." Further, it was directed that the Finance department should evaluate the project for out-of-pocket development cost. Turnbull reiterated that the V-8 project would be considered part of the sports car program, and any future plans were to be carried out only if it met future U.S. requirements.

The MG V-8 would be allowed to go ahead, effectively paving the way for the anticipated V-8 version of the Triumph Bullet. At the time it was predicted that the MGB line would be killed off, but including the V-8 for now could build interest and encourage a market for the newer sports car.

The project went ahead as planned and there was even a fanciful notion, allegedly the idea of

British Leyland decided to celebrate the fiftieth anniversary of Old Number One as the Golden Jubilee in 1975. Meant for the U.K. market alone, a limited edition of 750 special-liveried MGB GTs was announced, with black-and-gold painted Dunlop wheels similar to those of the contemporary MGB GT V-8. Note how factory shortages have led to one of the cars seen here at Abingdon being equipped with Rostyle wheels and whitewall tires. *Author archive*

Lord Stokes, to market the MGB V-8 in Japan. Although sales to the United States were supposedly the project's chief raison d'être (a prototype was even shown off to distributors there), in any event only the U.K. ever saw the factory cars, following their launch on August 15, 1973. They were received favorably, though the response was tempered by the engine's placement within the aging MGB GT platform that served as the basis for the GT V-8. (It's notable that no roadster was marketed.) The cars that saw the light of day offered little beyond the engine other than some fancy wheels and a handful of standardized trim items, most of which were already options on the cheaper 1.8-liter car.

Global conflict intervened to offset any buzz that had been generated. The conflict that broke out between Egypt and Israel in October 1973, known as the Yom Kippur War, plunged the world into a fresh fuel crisis such that, within weeks, demand for cars with 3.5-liter V-8 engines declined precipitously. No concession was made for the MGB GT V-8, which happened to be a surprisingly economical car that ran on cheaper, lower-octane fuel.

The MGB GT V-8 limped on for another three years, but after the first modest burst of enthusiasm, sales dipped toward unacceptable levels. Production ended quietly, much as the passing of the discontinued MGC seven years earlier. The basic formula was a good one, however: Private conversions continued, some of them produced by Ken Costello, and the concept would stage a renaissance in 1992, as we will see later.

John Thornley may have retired from MG by the time the MGB GT V-8 appeared, but he had strong views on how it was undermined

TOP RIGHT: This 1976 advertisement appeared at a time when sunroofs were becoming popular accessories. British Leyland cleverly pointed out that the MGB had the world's biggest sunroof. U.S. sales responded and went "through the roof." *Author archive*

RIGHT: A 1978 U.S. marketing photo of an MGB, complete with optional MGB side stripes. The latter were originally mocked up in the Longbridge design studio but ended up as a dealer-fit option in the United States. Before long, an open-top version of the Triumph TR7 would join the BL sports car family. *Marce Mayhew*

ABOVE: In 1979 it was announced that the MG Midget would be dropped from the range. Henceforth the Triumph Spitfire, whose 1,500 cc engine the Midget now shared, would cover the small sports car market. The last five hundred Midgets were painted black, with many going to Japan. This is the last car, at the factory on December 7, 1979. It is now at the museum in Gaydon. *Author archive*

by the forces at work in British Leyland. As he told the author, "The car retained the weight, balance, and behavioral characteristics of the B with power roughly equivalent to that of the MGC. It was a very, very good motorcar, so good in fact that it aroused a deal of jealousy among competing factions within BLMC. Production was constantly interrupted by engine shortages and some of the decisions as to where, or more particularly where not, it should be marketed could not be explained in normal terms."

RUBBER-BUMPER MGs

In North America, where markets were gripped by regulations in support of passive vehicle safety, there was a parallel increase of the influence by the insurance industry, which was frustrated by the spiraling costs of vehicle repairs and associated claims. It was their pressure as much as the growing demand for safety improvements that spawned the onset of new legislation aimed at securing zero-damage standards, calling for new cars to resist impact damage at speeds of up to 5 miles per hour (8 kilometers per hour) with no damage to critical systems such as lighting and fuel.

The new standards were phased in at a slower rate for imported sports cars than they were for domestics, but these had far-reaching impacts throughout all automotive markets. In 1973, MG Midgets and MGBs slated for U.S. markets saw a minor concession to this trend by sprouting large rubber overriders wrapped around substantial castings. They were known as Sabrinas, in honor of Norma Ann Sykes, a well-endowed British starlet who used "Sabrina" as her stage name. The following year, standards were tightened, necessitating new wraparound bumpers comprising polyurethane units with steel cores, fitted for all remaining markets where MGs were still sold. The MGB was fundamentally a strong car, so the conversion proved practical if costly. The Midget was a harder car to upgrade, however, and the exercise cost more, at a time when the engine of the smaller car was also switched to a Triumph 1500 cc unit shared with the Spitfire.

The polyurethane used for the bumpers was produced by Bayer, a German chemical company. While it could be produced in different colors, the MGs were produced only with the bumpers in black, molded by Marleyfoam. MG chief body engineer Jim O'Neill took out a patent for his design, while the exterior bumper shape came from the Longbridge studio. These resulted in what became known as rubber bumpers, though this was a misnomer. Although they were gradually accepted by enthusiasts, some of whom nowadays even prefer them to

the earlier chrome-bumpered cars, they were controversial at the time. The market soon recognized serious issues with the vehicles: poorer handling, a consequence of the raised suspension and the weight of the bumper armatures, and weak performance thanks to attempts to meet emissions rules.

In the midst of this, the MGB GT was withdrawn from U.S. sales in January 1975, in part thanks to the difficulty of meeting emissions standards due to its weight class. For similar reasons, it was said, the MGB was gradually removed from European markets in 1976. While emissions and safety standards were given as the reasons for stepping down the MGB's availability, surely there was no coincidence that this occurred around the time of the introduction of the Triumph TR7, the definitive Bullet corporate sports car, into those same markets. Even the TR7 saw modifications, however: Because of worries about coming legislation, it had been designed solely as a hardtop.

MISSTEPS ON THE WAY TO THE BETRAYAL OF ABINGDON

Although the original intention had been to develop an MG derivative of the TR7, it was also obvious that the home of MG would most likely not feature in this plan. The MG Midget segued in and out of the product plan, and several times it seemed likely to be euthanized (as it eventually was in 1979).

For a time other new cars were considered as potential bearers of the MG badge. This included, as we saw earlier, Condor, an MG coupe (ADO68) which was spun off, variously, to ADO28 (Marina), ADO67 (Allegro), or ADO14 (Maxi) platforms. After this path was abandoned, ideas emerged for an MG coupe that was derived from the proposed Marina replacement (ADO77); these also came to naught. Alan Edis, part of the Austin Morris product planning team at the time, told the author that "those ideas were little more than a footnote in the planning documents. They were never fundamental to the core program." Lastly there were thoughts of applying the MG badge to a version of the later Triumph Lynx, based on the definitive TR7, as a replacement for the MGB GT—another worthy idea that went nowhere. The simple truth lurking behind these proposals was that none of the concepts put forward would have relied on the Abingdon factory.

While the company was wracking its collective brain for new ideas, the U.S. market maintained its devotion to the MGB and Midget. In a market starved of genuine open-topped sports cars, the MGs, alongside the Triumph Spitfire, which generally outsold the Midget, found new friends, aided in no small part by clever marketing from Leonia's talented ad men—

OPPOSITE BOTTOM: In the immediate aftermath of the announcement by BL that MGB production was to cease, a consortium began to make plans to take over. This sketch by William Towns represents the proposed first-stage minor face-lift of the Aston Martin MGB. *Alan Curtis*

ABOVE: One part of the Aston Martin Lagonda Consortium's plans, this car represents the William Towns proposal for a modest face-lift for the MGB. It was rapidly turned into this one-off car, only shown to the press after the collapse of the venture in the summer of 1980. *Roger Stowers*

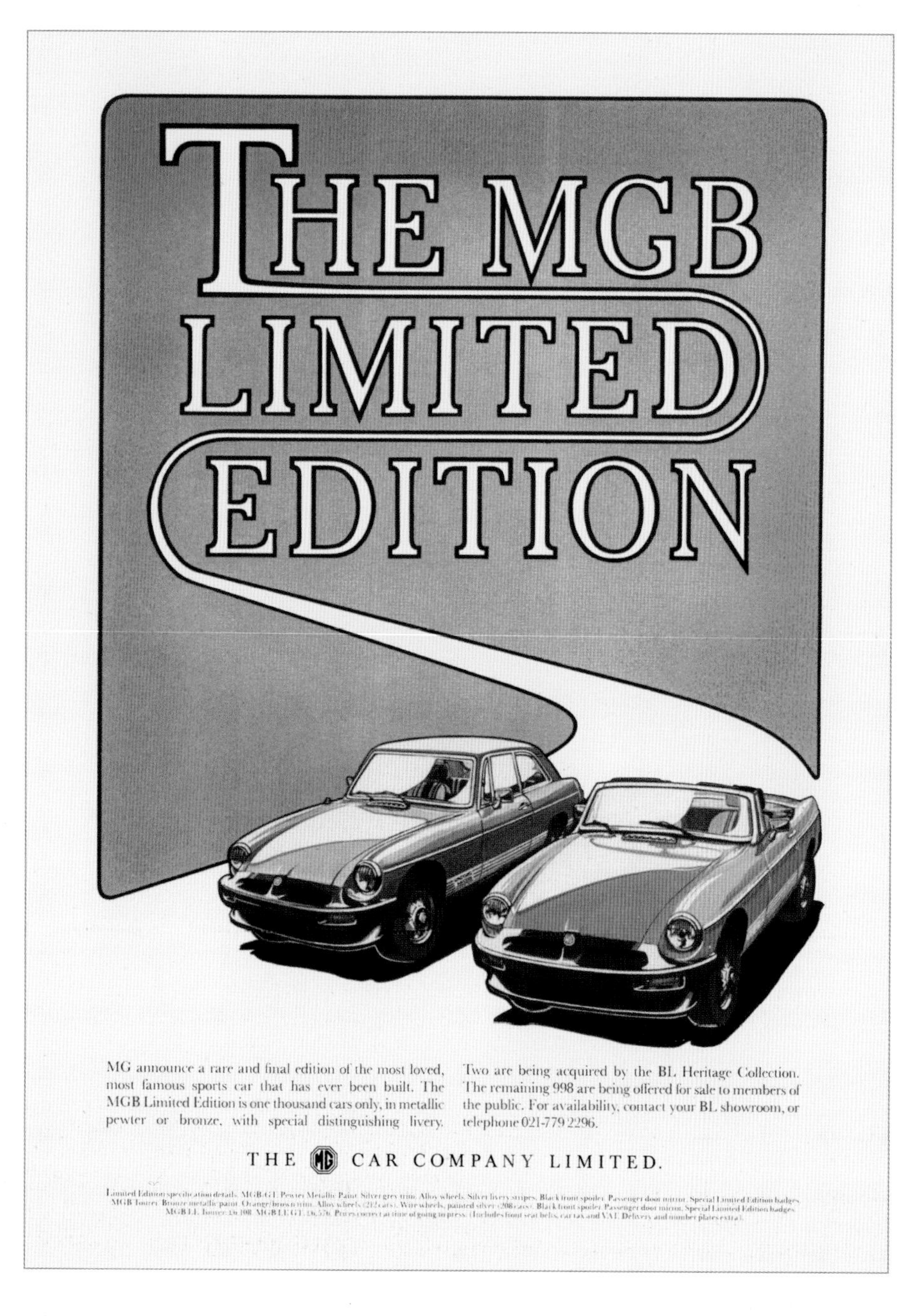

THE MGB
LIMITED
EDITION
MG announce a rare and final edition of the most loved, most famous sports car that has ever been built. The MGB Limited Edition is one thousand cars only, in metallic pewter or bronze, with special distinguishing livery.
Two are being acquired by the BL Heritage Collection. The remaining 998 are being offered for sale to members of the public. For availability, contact your BL showroom, or telephone 021-779 2296.
THE MG CAR COMPANY LIMITED.

The final U.K.-market Limited Edition MGB and MGB GT were built alongside regular cars in the final months of production, but they were released to the public only at the end of January 1981, three months after the factory had closed its doors. *Author archive*

and, for a while, favorable exchange rates. British sports car sales ballooned, and by 1977 MGB sales actually set a record in the United States.

This success in the United States was tempered by business woes back home: The financial collapse of British Leyland in 1974–1975, coupled with the inevitable focus on more important priorities, meant that MG sports cars never received the serious corporate attention they needed to compete. To establish needed stability, Michael Edwardes (later, from 1979, Sir Michael), a businessman hailing from South Africa who had managed several British companies over the previous decade, was appointed chief executive, then almost at once chairman, of the company in 1977. While the mainstream BL model range evolved, and new powertrains and other building blocks changed, the MGs became only more specialized and remote; the production of the parts for assembly became an inconvenience for the suppliers and source factories.

Things came to a head in May 1979, when the U.K.'s general election returned a Conservative government to be led by Margaret Thatcher. Likely a reaction to a recent period of industrial unrest and high inflation, the election result had far-reaching consequences, not least a shift in the dollar–sterling exchange rate, which veered sharply to the detriment of U.K. exports. Now known simply as British Leyland (BL), the company felt a direct impact from the new government. The new administration at 10 Downing Street, politically a world apart from that of the previous Labour government, trained its sights on the automotive conglomerate, with a focus on divesting it of its public ownership.

Export sales of an aged sports car range would have been a challenge in any event, but gas prices across the United States continued to rise, and further plans to regulate safety and emissions loomed large. In the midst of this emerged the long-expected open version of the TR7 and new V-8–powered TR8. At the time, fans of traditional sports cars often looked upon these Triumphs with disdain, including many Triumph as well as MG enthusiasts, but it was undeniable they were a more modern offering and arguably better suited for the future.

By the summer of 1979, BL was engaged in a process of cutting away the fat from its business, and the exchange rate was doing the poor MGB no favors. In September of that year, just as the town of Abingdon was concluding a weeklong celebration of fifty years of MG assembly in the town, the devastating news came that MGB production would be discontinued and the Abingdon factory turned over to different use. An international uproar followed, loud enough to reach the pages of the *New York Times*, which gave partial cover to an ultimately abortive bid by a consortium led by Aston Martin Lagonda (AML) to acquire the company. Their goal was to buy the factory, together with the rights to the MGB and the MG names.

Behind the scenes, work was undertaken to develop a low-budget face-lift of the Triumph TR7 to create an MG-badged version. Few thought such a conversion worth consideration, and it was widely seen as a desperate grab by BL for some credibility. Had it been built, it likely would not have saved Abingdon, though Swindon's Pressed Steel plant might have benefited. For a brief period after 1978, that plant had been responsible for the body shells of the MG Midget, MGB, and TR7, so it would have been the logical choice to manufacture the new car's body. When this plan was abandoned, BL planned a run-out limited edition of the MGB for the home market only, in roadster form (in Metallic Bronze) and GT (Metallic Pewter).

As it turned out, international finance woes scuppered the AML purchase. BL was planning to use Abingdon for other purposes, including complete knockdown (CKD) work to support Cowley and possible limited production of special versions of mainstream Austin Morris cars. These plans were dropped during the protracted talks with the consortium, and it became apparent that the factory would be sold off and most of its loyal workforce cast to the wolves.

Production at Abingdon finally ended in October 1980, with the U.K.-market LE models launched to the public in January 1981. These marked the end of a half century of car making in Abingdon and, for a short interlude, a continuity gap in the MG story.

ABOVE: In an inhospitable Detroit parking lot, BL chairman Graham Whitehead hands over the final U.S.-specification MGB (Limited Edition model) to Henry Ford II. The car was destined for the company museum. ***Author archive***

BELOW: The final Roadster, flanked by staff on the factory floor. The U.K.-market LE models were not made public until the following January, though some old hands from MG's past—such as Syd Enever and John Thornley—were invited in for a few sad photos, marking the end of fifty years of car making in Abingdon. ***Author archive***

MOTOR

WORLD EXCLUSIV

MG MIDGET

7

Changing Gear in the Eighties

AFTER ABINGDON

Abingdon's closure, and with it the end of the MGB, were met with incredulity by some, a sense of inevitability by others. Despite a last-ditch effort to salvage a low-budget joint MG and Triumph sports car out of them, the TR7 and TR8 followed the MGB into oblivion, and soon after a short-lived U.S. version of the Rover SD1 was ignominiously withdrawn. Within a few years, BL's once much-vaunted U.S. export program had receded to support Jaguar alone.

As far as many North American MG fans are concerned, this marked the end of the road for their beloved marque. Triumph fans in the same market were likewise plunged into sports car mourning after 1981. (The short-lived Triumph Acclaim, an early product of BL's deal with Honda, was never imported to the United States.) At the time of writing, North American fans of MG and Triumph sports cars have been without any new offerings for forty years and counting.

The first car produced in the post-Abingdon period came in the form of a small front-wheel-drive family hatchback. The product of BL's mass-market Light and Medium Cars Division, the Austin Mini Metro (the "Mini" prefix would soon be dropped) included variants with a choice of 998 cc or 1,275 cc improved A-Plus engines. Support for the new car, which included a major refit of the Longbridge

OPPOSITE: Created by Gerry McGovern in 1984, this MG sports car concept was never shown to the public. Its existence was leaked in a (largely inaccurate) article in the U.K. magazine *Motor*. The car was conceived as a potential offshoot of the proposed R6 Metro replacement, though no engineering platform work was ever undertaken. As we shall see, McGovern went on to oversee what became the MG*F*. Today he is design director at JLR. *Author archive*

Some saw the new MG Metro 1300 of May 1982 as a cynical ploy to capitalize on the emotions generated by the end of MG sports car production at Abingdon. Competent and appealing, the small car won many friends anyway, along with the backing of both of the main U.K. MG clubs. *Author archive*

plant, involved a massive investment of £275 million (around $650 million at the October 1980 exchange rates) and a patriotic launch campaign. The new Austin Metro was positively received and sold well, some two million of all variants over its lifetime; an early customer was Lady Diana Spencer, the future Princess of Wales.

By this time British Leyland had adopted a new name for its production division, Austin Rover Group. One of its first acts, following on the MGB and TR7 debacles, was to develop a high-performance derivative (HPD) of its new Metro and an even more powerful, very high-performance derivative (VHPD), the latter slated to receive turbocharging as a more viable alternative to a bigger powertrain. These additional models appeared in 1982: the MG Metro 1300 in May, followed a few months later by the MG Metro Turbo.

Although these new MG-badged models initially sparked mixed emotions, they were welcomed by the two main U.K. MG clubs with open arms. A clever motorsports provenance was carefully devised, with the Metros participating in both MG club and wider race series. European exports soon followed—but these models were denied a North American audience.

Sometime after these events, the author interviewed Harold Musgrove, Austin Rover's managing director and later chairman, asking him specifically about the contemporary clamor for a new sports car in the years following the death of the MG and Triumph ranges. According to Musgrove,

> *A sports car with anything other than an interior structure coming off a mass-produced base would be impracticable. People would say "why not do a one-off of a Metro—it would be lovely?" but they didn't know the practicalities. Things like the four-speed gearbox, the monosides all done in a single pressing . . . this kind of thing depends strongly upon how profitable you are—and the Government was a shareholder. MG was part of the corporate plan, but it would have come off the relationship with Honda. We wanted to do a sports car but one that was based on a volume car.*

One is tempted to comment that nobody at Mazda seems to have felt so constrained. After the LC8 Metro came the LC10 Maestro. Musgrove told the author that this was where he became concerned: "LC10 was the problem. We had got the Metro going—highly profitable, particularly the MG Metro—and we were starting to break through. I was now the managing director and I took one look at

the Maestro, and it was tooled and ready—but I said to Ray Horrocks [his boss] that we needed a new styling director."

David Bache was the former Rover styling director who had seen through such vehicles as the Rover P6, Range Rover, and Rover SD1 3500. Now elevated to Leyland's overall director of design, his service under Musgrove as joint Austin Rover director rapidly became untenable. When Bache left the company, therefore, and British ex-pat and former Chrysler designer Roy Axe was recruited, it was too late to alter the Maestro in any meaningful way. Enhancing the related LM11 saloon as the Austin and MG Montegos (including the blisteringly fast MG Montego Turbo of April 1985) seemed a way forward. As Musgrove said, "Roy Axe appeared. I gave him the Montego. We could hardly survive on Allegro, Princess, Marina, and Maxi—we just *had* to go for Maestro. The engines in the original plan had been A-Series with O-Series to follow after 18 months. Then someone said that we could do something with the E-Series—so we went for that in a hurry."

The MG Maestro certainly exhibited many of the hallmarks of a project that had, as Musgrove put it, been hurried. Its initial 1.6-liter guise featured twin Weber carburetors, which remarkably saw two different engines over its short life—the R- and S-Series. These frequently suffered fuel vapor lock that could render the car difficult to restart from hot. Within a short time, though, the much better two-liter fuel-injected MG Maestro arrived, offering a capable high-performance hatchback (dubbed a "hot hatch") to rival its competitors, the Golf, Escort, and Astra. Ironically, in a different universe, that fuel-injected O-Series engine would have seen prior service in enhanced versions of the MGB and TR7.

In due course, a bold experiment to create an Austin Rover rally weapon also bore fruit in the form of the mid-engined MG Metro 6R4. There was also a sleek concept car, the MG EX-E, packaged around the V-6 running gear of the 6R4; this model stunned the Frankfurt Motor Show audiences when it debuted on September 10, 1985.

ABOVE: The MG Metro Turbo was developed with input from Lotus. Early prototypes were potent and fast but tended to be fragile. Pictured is an all-white version of the face-lifted model from 1986. *Author archive*

LEFT: A face that only a mother could love? The styling of the Maestro was more or less settled in 1976, but the production car—the Miracle Maestro—didn't appear until March 1983. The original MG version had a problematic Weber equipped 1.6-liter R-Series four-cylinder engine. *Author archive*

Musgrove vacillated over revealing the EX-E to the wider world, but in the end he relented. As he told the author, "The styling basis of EX-E was intended to lift the MG image. The culture of MG had been lost—there had been no new designs for years—so Roy Axe was aiming to rebuild the culture, and in the United States in particular, you need an image."

Roy Axe was justifiably proud of the EX-E concept, which showed the world that, Maestro and Montego notwithstanding, Austin Rover now had a world-class design studio. But with the government seeking to divest itself of the business, there was nervousness about what some might have seen as a frivolous project to put into production. Axe told the author: "Now the EX-E, you see: That's a car that could have been produced as purely a low-volume very prestigious thing that would have lifted the image of Austin Rover when it really needed lifting, but nobody had the courage to really do that."

It needn't have gone this way: Beyond the EX-E, the design studio under Axe had developed a striking proposal for a new open-topped sports car, the MG F-16. The author spoke to Axe about this some years later.

> *They did a lot of research on MG, and Marketing would always insist that MG was a "brand of the past"—it was not a brand that the modern man or woman related to. They thought that it related to something like the MG TC Midget in*

TOP LEFT: The superior face-lifted version of the MG Maestro. The two-liter O-Series engine in fuel-injected form made for a fast and fairly refined five-door hatch, with room for the entire family. This is a 1988 specimen in Metallic British Racing Green. Although exported to Europe and in small quantities to Japan, the Maestro never made it to North America. *Author archive*

MIDDLE LEFT: Like the MG Metro Turbo, the MG Montego Turbo was offered for a time in an all-white finish. Note the smooth aerodynamic alloy wheels, very much a product of their time on this 1986 car. *Author archive*

LEFT: A later MG Montego Turbo, finished in the popular new color of Metallic British Racing Green, with spoked-pattern silver alloy wheels. *Author archive*

terms of how the Americans saw it—which was quite right, as they did. But what they totally miscalculated with regard to America was that the Americans had their image of MG, but because of that, MG was on a pedestal, and you could have brought a modern MG out and it would have absolutely wiped the floor. And the car that we produced as a show car, F-16—well, it was never shown, actually.

Axe certainly felt that this was a lost opportunity.

Now that could have been in production eighteen months ahead of Mazda's Miata: Everything was there—all the data was available that the Mazda people were also looking at—and I'm sure that the market research people said, "It'll be a hard battle—very difficult. . . ." and so on, and of course the argument over here in England was "Well, very nice—but the only market we've got is England, because the Americans will never buy an MG, so there's no point in putting it over there, and therefore the British market volume would never sustain the car." They did that over and over.

Thus there were certainly some tentative ideas for a brand-new MG sports car, or even a coupe. Austin Rover, recast as Rover Group in 1986, had more pressing concerns. The 1980s remained a sterile decade for open-topped MG sports car fans, unless one counts the handful of somewhat dubious Metro conversions trotted out in the U.K. by some enterprising aftermarket companies.

An interesting addition to the range for the 1989 model year was a turbocharged version of the MG Maestro, which effectively married the powertrain of the Montego Turbo with the lighter Maestro bodyshell. It was dressed up with new exterior moldings and built in a very limited run of 505 by Tickford Engineering; 215 examples were painted Flame Red, 149 Metallic British Racing Green, 92 Diamond White, and 49 Black.

The fact that launch publicity referred to this new model simply as an MG Turbo perhaps hints at the fact that Rover Group management, cementing its rebrand by openly eliminating the Austin name, was at the same time acutely conscious of the value of the MG name. As Musgrove had noted, some of MG's culture—for which read "credibility"—had been lost.

Performance of the turbocharged MG Maestro was certainly impressive for a

The MG Metro 6R4 was a brilliant barnstorming rally weapon, with a bespoke, naturally aspirated, midmounted V-6 engine and four-wheel drive. Sadly, the rally class it was designed for was soon canceled in the wake of a tragic incident (unrelated to the MG) on the Tour of Corsica in 1986. *Author archive*

ABOVE: Photographed in the secluded Canley Design Studio viewing garden, this is a full-sized concept model for a possible new "MG Midget" that might have been based on a future mainstream small car. It never reached production, even if a "leak" led to a front cover story on one of the U.K.'s leading weekly motoring magazines (see page 176). *Author archive*

roomy five-seater, five-door hatchback of the era. With a claimed 0–60-mile-per-hour (0–96.6-kilometer-per-hour) time of just 6.7 seconds and a top speed of 128 miles per hour (206 kilometers per hour), it was the fastest production MG at that time.

PRIVATIZATION: BRITISH AEROSPACE

A long-standing ambition of the Conservative government under Margaret Thatcher had been to return the company to private ownership. Having undergone a metamorphosis from British Leyland, through Leyland, to BL, to Austin Rover, and then Rover Group (and meanwhile forged a productive relationship with Honda of Japan), the company by any other name was still being supported by public funds throughout Thatcher's tenure as prime minister. As an avowed champion of private rather than public sector ownership, she presided over the privatization of many well-known parts of public life—clearly, she wanted to see still more such changes. In this context, notwithstanding what she saw as the comparative success of the now-departed Sir Michael Edwardes's battles with the unions, she was determined to see Rover Group off-loaded.

Talks in 1985 and 1986 with various motor industry suitors, including Ford and General Motors, were abandoned mostly for political reasons. The eventual answer to the prime minister's prayers came with the sale of Rover Group to British Aerospace (BAe), an unlikely choice, in March 1988. The deal was designed to protect as far as possible the relationship forged with Honda, and there were vague but mostly unconvincing words about "synergy." These were perhaps intended to draw eyes away from some of the sale's underlying motives, not least the significant opportunities dispose of or re-develop some of Rover Group's ample land assets, via a BAe subsidiary called Arlington Securities.

RIGHT: Austin Rover designers came up with a possible new generation of MG sports car, known as the F-16 in honor of the fighter aircraft. This concept was being produced around the same time that Mazda was working on what became the Miata for 1989. *Author archive*

In the run-up to the BAe sale, there had been some realignment of Rover Group management. A purge saw several familiar faces, including Chairman Harold Musgrove, leave the business, and Sir Judson Graham Day assumed the new role of company leader. Known as Graham Day before being knighted in 1989, he was a Canadian who had previously come to public attention through his leadership from 1984 of British Shipbuilders. Day arrived at Rover Group, saw through the BAe deal, and remained as chairman until 1991. He sought to encourage the renaissance of MG, declaring as much publicly on a number of occasions. It was only after he left the company, however, that real steps in that direction were made.

BRITISH MOTOR HERITAGE AND THE MGB

By the time the MGB was discontinued, Pressed Steel's plant at Swindon had assumed responsibility for the manufacture of the TR7 body shells, and thus for a brief period was something of a specialist sports car body assembly center. Although tooling for the MGB was decommissioned, a supply of spare parts such as wings and doors remained in production, as these continued to be service requirements for years into the future.

The bulk of the tooling related to the bodywork itself was carefully wrapped and stored outside the factory; the hope was that, one day, there might be demand to do something more with it. This was remarkable, as the tooling was often sold for scrap and melted down for its metal value—as indeed happened with some of the TR7 press tools.

The person who preserved the MGB tooling at Swindon was David Nicholas of Pressed Steel. A few years later it would be another David—David Bishop, from the Body Engineering team at Cowley—who would take up the baton and effectively bring the MGB bodyshell back to life.

More often seen in silver, the unique MG EX-E was briefly finished in red, as it appears here in the Austin Rover Design viewing garden at Canley. *Author archive*

ABOVE: The MG Maestro Turbo was a low-volume offering that used the services of Tickford Engineering as part of the build process. At the time it was the fastest-ever production MG. *Author archive*

BELOW: April 1988 saw the launch of the MGB bodyshell at British Aerospace HQ in Central London. Here British Motor Heritage's Peter Mitchell *(left)* sits next to David Bishop, the man chiefly responsible for bringing the MGB bodyshell to fruition. Behind them is Old Number One itself. *Author archive*

Bishop had joined the British Motor Heritage offshoot of the British Motor Industry Heritage Trust, a charitable body that had begun as a Leyland initiative aimed at saving and securing the legacy of the many constituent companies subsumed by the main business. Over the Christmas break in 1986–1987, Bishop settled down with the 250-page document that listed all the MGB body tools, which aided him in cross-referencing what was needed with what had survived.

In the spring of 1987, paper study gave way to a comprehensive search of what remained at Swindon and other stores held by external suppliers. Meanwhile, premises were secured at Pioneer Road, Faringdon, Oxfordshire; like Abingdon, it was situated in the Vale of the White Horse and thus not far from the various factories that had once supported MGB assembly. Setting up at this location would mean that the new units produced in Pioneer Road were near the homes of retired plant personnel—the

Prime Minister Margaret Thatcher, as head of Britain's Conservative Party, was no fan of the public ownership of MG's parent business she had inherited from the previous administration. Here she is in Downing Street, London, seated in a Naylor TF 1700. This replica of the old MG *TF* was aimed at recapturing some of the appeal of the older MG sports cars. *Author archive*

people David Bishop would ask to come back and take part in his new venture.

"There were some eight hundred press tools in all, weighing over one thousand tons—which meant fifty lorry loads to be moved to the operation we set up at Faringdon," Bishop told the author at the time. Some of these tools had to be repaired on location, while others were dealt with at the pressing plant. Remarkably, only four dies were found to be missing and had to be remade, while other significant finds included framing jigs from the former CKD operation at Booth & Poole in Ireland, where modest numbers of MGBs had been assembled; these were invaluable for the low-volume production being planned.

The project was unveiled in April 1988, just days after the aforementioned British Aerospace deal, and in short order both right-hand- and left-hand-drive MGBs were rebuilt for use as publicity tools. The first RHD MGB became a part of the vehicle collection now held at Gaydon, while the LHD specimen was driven across the United States as part of a magnificent publicity exercise.

The factory premises later supported the MG RV8. As the business grew, production was shifted to a new base at Witney, which continues to build many bodies, such as the TR6, Mini, and Midget bodies, along with other projects. It is a sobering to think that, at the time of writing, British Motor Heritage have been building MGB body shells for longer than the cars' original production run from 1962 to 1980.

GRAND PRIX
DUNLOP
MG

8

The Nineties: Return of the MG Sports Car

HANDOVER PERIOD

The early 1990s saw the beginning of a transition for the Rover Group as the Metro, Maestro, and Montego ranges came toward the end of their respective roads. With the end of the Austin name, the portfolio of marque names shrank to a choice of the various Rover brands (including Land Rover and Range Rover). As far as the public was concerned, there seemed to be an uncertain future for MG. When a Metro GTa appeared, offering many of the attributes of the MG Metro 1300, some observers began to ask what was to become of the MG models. The situation was not unlike what had happened in 1969–1970, when the Austin and Morris 1300 GT models had elbowed out the MG and Riley 1300 models. Even so, it is worth pausing to acknowledge the good business that the MG saloons and hatchbacks had notched up by the time they finished.

Shortly before the decade began, Sir Graham Day asked senior managers in Rover Group if they could deliver a proper MG sports car for a modest budget. One of the designers, Richard Hamblin, decided to see if it could be done. He initiated a study that evolved into a wider body of work that embraced ideas such as the return of the classic Mini Cooper and three parallel MG sports car roadster studies. The MG ideas were, in turn, a front-wheel-drive car (notionally based on the Maestro or,

OPPOSITE: A special sports-racing version of the MG*F* was conceived jointly by Rover Group's design team and body contractor Mayflower. The idea was to use a supercharged version of the K-Series engine in a subtly widened body shape. The result was called the MG*F* Super Sports, bringing back a name from MG's earliest days. First shown at the 1998 Geneva Salon, the car was also allocated an EX number (EX254). *Author archive*

MODEL	UNITS PRODUCED
MG Metro	120,197
MG Metro Turbo	21,968
MG Maestro 1600 R	12,398
MG Maestro 1600 S	2,762
MG Maestro EFi/2.0i	27,880
MG Montego EFi/2.0i	37,476
MG Maestro Turbo	505
MG Montego Turbo	7,276
TOTAL	**230,462**

The MG Metro 1300 proved a popular hatchback with sporty overtones: a good first car for many, not unlike the much later MG 3. This is a 1989 Model Year specimen. ***Author archive***

potentially, the future AR8 Rover), a more traditional front inline engine with rear-wheel drive (which would become a Rover V-8–powered concept), and a mid-engined sports car that used a Metro drivetrain and subframes (echoes, perhaps, of the ADO21?).

These concepts were quickly wrapped up in a project overseen by a new group called Rover Special Products (RSP), who dubbed the MG project "Phoenix Revival." The aforementioned trio of individual themes became PR1, PR2, and PR3. In time there would be two more added to the list: PR4, or "Adder," an offshoot of the MGB Heritage bodyshell venture; and PR5, a more upmarket sports car that could have been either a traditional V-8, like PR2, or a concept evolved from the still-current Rover 800, sold in North America as the Sterling.

The culmination of these strands of MG sports car concepts saw a number of retirements and scrapped plans. These included the end of the MG Metro, Maestro, and Montego lines and the abandonment of a tentative plan to apply the MG badge to Rover's self-engineered offshoots of the important new Rover R8 200 range.

Code-named "Tomcat," the coupe version of the Rover R8 arrived in October 1992, at the same show that saw the launch of the MG RV8. Effectively a restyled and upgraded MGB "evolution" with a Land Rover 3.9-liter fuel-injected V-8, the RV8 featured smart leather-and-burr-elm interior trim, with a price tag roughly equivalent to a well-specified Rover 800 of the time. Initial sales of the RV8 were sluggish after the initial burst of enthusiasm (a recession in the U.K. did not help), but the new retro MG was well received a year later at the Tokyo Motor Show. It went on to sell all of its planned small production run, albeit over a four-year period.

More vital to the renaissance of MG was the 1995 MG*F*, with its italicized *F* deemed an element of the model name. This neatly styled, cleverly engineered mid-engined roadster offered a choice of Rover's own award-winning K-Series engines. As an echo from MG's past, the new enlarged engine capacity chosen for the MG*F*'s K-Series was 1,796 cc, almost the same size as the MGB's 1,798 cc B-Series of 1962. In much the same way as that earlier engine, the B- and K-Series engines had both begun with much smaller capacities but had expanded with model need.

The MG*F* was launched at the March 1995 Geneva Salon. The author joined a small party of MG Car Club members who drove all the way from Abingdon to welcome the new sports car's debut in Switzerland. In the period between the MG*F*'s conception and its launch, however, there had been yet another palace revolution.

ARRIVAL OF THE BAVARIANS: BMW BUYS THE ROVER GROUP

Germany's BMW was expanding in the early 1990s. The company was looking for a way to expand its portfolio to embrace the higher-volume sectors covered by bigger rivals like VW AG (including the Audi brand) as well as the anticipated growth of the larger 4x4 sector, the most famous of which were the Land Rover and Range Rover brands. British Aerospace had almost exhausted its sell-off of surplus assets by this time and now found itself needing to direct its diminishing cash reserves toward other priorities. When BMW came calling, therefore, the British aerospace giant was ready to sell. Honda offered shared ownership as a counteroffer, but the Bavarians' cash was too tempting.

In the eyes of many within Rover Group, this was a poor way to treat the Japanese company whose collaboration had been so instrumental to its recovery, but ultimately financial equations tend to dominate such decisions. BMW expressed a willingness to maintain the relationship between Rover and Honda, but this was unlikely: From Honda's perspective, BMW was a key and powerful industry rival that, moreover, had barged into what it saw as a settled relationship.

Because the MG*F* was a mid-engined sports car derived in some way from parts of the then-current Metro sedan, with very little if any Honda input, BMW was sanguine about letting it continue. Within months, BMW AG began

ABOVE: A coupe version of Rover's AR8 joint venture with Honda was part of the plan from early on. This concept wears MG badging, but when it was decided to reserve the octagon for "proper sports cars," the model range was developed as an entirely Rover-badged family, including both a convertible and a coupe. *Author archive*

BELOW: The Phoenix prototype PR1 was closest in size and style to the F-16 master bodyshell on which it was based. Built in steel by contractor Motor Panels, it used an MG Maestro front-wheel-drive platform. Had it been built, it likely would have used Rover R8 components). *Author archive*

LEFT: The PR2 was built on a ladder platform chassis, with a front-mounted in-line Rover V-8 engine driving through the rear axle. The prototype was built by Reliant and had a beautifully finished composite bodyshell. *Author archive*

BELOW: The PR3 was dubbed the "Pocket Rocket." It used Metro running gear mounted behind the seats. Such an idea echoes the MG ADO 21 we saw in chapter 6. *Author archive*

production of a new, more traditionally arrayed roadster, the Z3, which was built in a brand-new factory in Spartanburg, South Carolina. The outcome might have been quite different for the MG*F* had the model been targeted to the U.S. market, and not tied to a financial partnership with Mayflower of Coventry, its bodyshell supplier. All seemed well for MG under the BMW umbrella for a while, though in the meantime, unsurprisingly Honda took umbrage and began to withdraw from the wider Rover relationship.

Like MG, Range Rover and Land Rover were more nearly immune from this, as they had for many years been separate from the mainstream Austin, MG, Triumph, and Rover lines. BMW chairman Bernd Pischetsrieder was related to Sir Alec Issigonis and a self-professed Anglophile.

After being photographed alongside an MG RV8 with Rover Group chairman John Towers, the head of BMW was quoted in an American car magazine interview as suggesting that "he could see a V-8 in MG's future," a message that seemed encouraging to anyone concerned with how the British sports car marque might sit in a family where the BMW brand was marketed as the "Ultimate Driving Machine."

The MG*F* went on to form the basis of popular and successful race series in the U.K., France, and Japan, although frustratingly for North American MG enthusiasts, this "English mini-Ferrari" was never shipped across the Atlantic. In the MG*F*'s home market, sales took off sufficiently well that the figures consistently exceeded those of the Mazda MX-5, as the Miata was known outside North America. From time to time there were rumors either that the MG*F* would be reengineered for U.S. sales, or that a new MG might spring from the BMW Z3 platform when the BMW-brand sports car moved on to a new generation Z4 model. All we saw in practice, previewed at the Geneva and Frankfurt shows, was a concept for an MG*F* with slightly wider wings and a supercharged version of the K-Series engine. The concept was evocatively named the MG*F* Supersports, harking back to one of the early MG model names from the 1920s.

All was not well in the Bavarian boardroom, however: Although the MG*F* was selling well in such markets as the U.K., Europe, Australia, and Japan, the real problems were with the rest of the range, now that Honda had largely walked away from the relationship and Rover was reliant upon a range of mostly Honda-based hatchbacks and sedans. Initial ideas from the Rover side were for badge-engineered BMWs—after all, hadn't Rover and its forebears long followed such thinking successfully, most recently with their Honda-based models?—but these were soon swept aside in favor of a direction from Munich for Rover to develop its own platforms.

Board member Wolfgang Reitzle was one of BMW's fiercer critics of the relationship and someone whose interest in Land Rover (but comparative hostility toward the Rover cars side) was well known. In the recent past, prior to BMW's purchase of Rover Group, Pischetsrieder and his colleague had been viewed as rival candidates for the BMW chairmanship. Indeed, Reitzle had led the project to create the new Spartanburg factory as a BMW bridgehead in North America, but he had lost out for the top job in Munich's distinctive BMW HQ. Reitzle had been less convinced of the merits of the Rover acquisition and had even publicly criticized the quality of the Land Rover and Range Rover model ranges.

BMW kept their Honda-based Rover cars. The second-generation HH-R and Theta Rover 400 of 1995, among others, were too close to production; abandoning them at this point would have been irrational. The Bavarians' major focus, however, seemed to be on a new superb, though somewhat retro, walnut-and-leather–lined Rover 75 midrange saloon, intended to be BMW's very British answer to the Audi A4. John Towers stepped away from his role as chairman of Rover Group in 1996, his place taken initially by Wolfgang Reitzle and then, on April 30, 1997, by BMW man Dr. Walter Hasselkus. Some years earlier, Hasselkus had headed BMW U.K. Like Pischetsrieder, he was considered a warm character and an Anglophile. Some of his BMW colleagues jokingly referred to him as "Sir Walter."

By this stage, however, there were wider market issues to contend with. The U.K. had refused to adopt the new European joint currency, the Euro, or to embrace the EU's exchange rate mechanism, which would have placed a check on the ability of the pound sterling to move against other European currencies. (Other countries had also proved slow to accept the Euro, including Germany, which gave up its Deutsche Mark in January 2002.) In a curious way there were some parallels to the exchange rate issues that had hastened the demise of the MGB, but in this case on a more critical companywide basis.

BELOW: A bigger sports car rounded out the series: The PR5/DR2's origins lay in an earlier design exercise called DR2. The prototype was built on a second-hand TVR. A larger sports car of this type was considered a safe bet for U.S. sales alongside the Sterling, which was available from 1987 until that car was withdrawn in 1991. *Author archive*

BOTTOM: Nicknamed the "Adder" as a nod to the classic Shelby/AC Cobra, the PR4 project was a clever marriage of the Heritage MGB bodyshell, reengineered components, and some subtle restyling. Launched in 1992 as the MG RV8, this was a limited-production declaration that the MG sports car had returned. Before the arrival of BMW in 1994, there was even a plan to sell off the entire MG project, with the RV8 held up to sweeten the deal. *Author archive*

BMW had bought currency in advance as a way to help protect its business, but all such schemes eventually run out of road. Now critical voices were being heard discussing what was to be done with the company's British division, described by German media and some within BMW as their "English Patient" in a play on the name of a popular contemporary movie. At the Frankfurt Motor Show in 1997, an advanced glimpse of the new Mini promised for 2000 was unveiled, but despite that note of optimism, Hasselkus admitted that BMW's plans for Rover were not going according to plan.

Politics even managed to overshadow the launch of the Rover 75 in October 1998, a date that had been moved up, at Pischetsrieder's insistence, to deliberately clash with Jaguar's unveiling of their new S-Type. The new Rover was unveiled in a grand manner on the Rover stand. Less than an hour later, in a BMW press conference on the other side of the National

Exhibition Centre, the BMW chairman was complaining about the financial problems that threatened to undermine further investment in other new projects. The author was in the audience, a lone voice asking about U.S. sports car plans; "Walter has an Austin-Healey in his garage" was all that the chairman would comment, with a laugh.

On December 2, 1998, Walter Hasselkus somewhat unexpectedly fell on his sword, shouldering blame for Rover's continuing losses. His departure was viewed with some regret in the U.K., but as ever in big business, someone often has to be sacrificed to satisfy the markets. Hasselkus was replaced by the somewhat lower-profile Werner Sämann, described as "an engineer and academic," and also rather less obviously Anglophilic than his predecessor. The change came as Rover Group shop stewards agreed to a radical cost-cutting plan aimed at saving £150 million sterling ($247 million) a year before 2000, long projected as the time when a turning point would be necessary.

The next stage of the "English Patient" drama unfolded on February 5, 1999, when both Bernd Pischetsrieder and Wolfgang Reitzle were forced to resign from BMW. "BP" was heavily criticized for the whole Rover Group affair, although he nevertheless departed with a £6 million golden handshake, and a year later was in the top seat at VW AG. Meanwhile Reitzle's supposed plans—to close Longbridge and switch Mini production to Cowley, Oxford, where the upmarket new R75 model was being built, and keep Land Rover at Solihull—were rejected. His ideas were ultimately not that far off the mark.

OPPOSITE TOP: The Heritage bodyshell factory, then based at Faringdon, Oxfordshire, was responsible for building the bodies for the MG RV8. The bare metal "body in white" units were then shipped to Cowley for painting and subsequent assembly. *Author archive*

OPPOSITE BOTTOM: With the RV8, MG production was back at Oxford for a brief period. This was a homecoming of sorts, with manufacture returning to where the MG story began, even predating the fifty years of Abingdon assembly. This is an early Flame Red MG RV8 on the production line at Cowley. *Author archive*

LEFT: The author's own Nightfire Red MG RV8 reflecting the clouds. The RV8 was never a big seller in its home market, but export sales to Japan helped bolster the company's finances. Many RV8s, including this one, eventually returned to the U.K. from Japan. *Author archive*

ABOVE: The rear view of the MG RV8 shows the subtle flares and curves that were introduced to beef up the MGB body shape. The project code for the car shared its name with the PR4: "Adder." The taillamps were entirely bespoke for this car alone. *Author archive*

BELOW: The fascia of the MG RV8 is far more luxurious than the MGB ever was. The wood is burr elm. *Author arhive*

Return of the EX Numbers

Even if BMW outwardly demonstrated little obvious interest in MG, within Rover Group there was an appetite to try to recapture some of the glory of MG's past record-breaking history. The great MG record breakers had been known by their project codes from MG's internal EX Register, which began with EX101. Its roll of famous models included EX120, EX127, EX135, EX179, EX181, EX182, and so on. The list was kept in the MG design office, and the master copy thankfully survives at Gaydon. The author had long had his own copy when Rover Group's Kevin Jones got in touch and explained what was in the works: A record breaker was being planned, loosely based on the MG*F*, and this was initially known as EX*F*. Jones knew that the name could only serve for the one car; however, he wanted to know how high the old Abingdon list had got.

With both Pischetsrieder and Reitzle gone, the BMW chairmanship passed to another board member, Professor Joachim Milberg. In July 1999, Milberg announced that the Rover 200 and 400 would be replaced by a single medium-sized car, to compete against the Ford Focus and Vauxhall Astra, which was to be built at Longbridge. "The new medium car will have a similar wide variety of derivatives as the BMW 3-Series," he told the media, "to develop a whole family of cars from which we expect significant growth in volume." BMW was belatedly playing catch-up with the volumes it needed from Rover to remain viable, the figure projected by the company being for Rover to make $1.8 to $2 million annually to remain competitive.

Throughout this volatile period, nothing was said about the prospects of creating a new sports car to take over from the MG*F* and perhaps reenter the U.S. market. Once again, the MG brand was seen as some way down the corporate ladder of importance.

It so happened that EX250 had been the project code number in the register for a one-

The MG*F* was unveiled to the author and a handful of club officials before its public launch at the 1995 Geneva Salon. Note the presence of Old Number One and an MG TC Midget. Rover Group was understandably keen to emphasize the true sports car provenance of this new model. *Author archive*

BELOW: Terry Kilbourne, a technician from a California Land Rover dealership, had the necessary SCTA license and prior experience to drive EX*F*, which featured a much-modified 1.4-liter Rover K-Series powertrain. *Author archive*

OPPOSITE TOP: EX256 (see table at right) was a striking concept for a possible supercharged version of the MGF, the so-called "MGF Super Sports", shown at various European motor shows, but sadly destined never to see production, a victim in part of the break between Rover Group and BMW. *Author archive*

EX VEHICLES OF THE LATE 1990S

EX CODE	PROJECT	DATES
EX253	MG EX*F* 1.4-liter record breaker for Bonneville, 1997	August 1997
EX254	MG Super Sports concept car, Geneva 1998	March 1998
EX255	MG record breaker for Bonneville, 1998	August 1998
EX256	MG MG*F* Super Sports roadgoing concept car: first unveiled at Geneva in 1999, finished in black with red leather interior (seen again at the MG*F* 2000 Model Year launch at Goodwood in July 1999), then refinished in Pearlescent "flip-flop" Chromaflair gold/green with green leather interior for the September 1999 Frankfurt Motor Show	March 1999

off safety vehicle, based on the MGB GT, known more commonly as SSV-1. EX251 and EX252 had been rather less exotic projects related to the Austin Mini, Allegro, and Princess, which meant that EX253 was in theory available. EX*F* therefore became EX253, and other projects listed in the table above appeared over the following two years. As we shall see in the next chapter, the list continued to evolve.

Afterward, Nick Stephenson, Rover Group's design and engineering director, stated that "it is fitting that they should be rewarded with the knowledge that they have produced the fastest MG ever. My colleagues on the Rover Group Board gave the project their blessing in June, and we are delighted to see the team exceed their target of 200 miles per hour (322 kilometers per hour) in such a short period of time. They cannot wait to return to the salt next year with their sights firmly set on the records of Stirling Moss and Phil Hill."

EX*F* was not the vehicle to answer that challenge, and consequently EX255 was created, a bigger project with a twin-supercharger equipped much-modified Rover V-8 designed to deliver around 900 brake horsepower, with the driving by land speed record-breaker Andy Green. Sadly, the car ran into engine difficulties in the garage on site, and never ran on the salt. By this time BMW was tightening its purse strings, so the project was abandoned.

The Record Breakers Return

For a brief interlude in the BMW years, when Rover Group was able to indulge in activities hitherto unthinkable, there was a modest resurgence of record-breaking fever. Both the projects drew heavily on the platform of the MG*F*, although the resemblance was mostly visual. The EX*F* was run first, driven by Terry Kilbourne, a technician from a Californian Land Rover dealer. He had been chosen because he held the necessary Southern California Timing Association (SCTA) competition license. The 1.4-liter-engine EX*F* achieved a respectable figure of 217.4 miles per hour (349.87 kilometers per hour) at Bonneville on August 20, 1997.

LEFT: Preparation of the EX255 record breaker at Gaydon. A serious effort, in the event it was stymied by the failure of a part that couldn't be replaced on the ground in Utah. The planned driver was celebrated record-breaker Andy Green. Sadly, it was not to be, and other problems with the BMW ownership of Rover Group meant abandoning a plan to return the next year. *Author archive*

MG TF
MG ZR
MG ZS
MG ZT
MG

9

Phoenix Rising: Into the Millennium

THE END OF THE BAVARIAN LOVE AFFAIR

As we saw in the last chapter, the rancor that eventually resulted in the "divorce" between BMW and Rover Group had been growing long before 1999 rolled into the new millennium. Over the winter of 1999–2000, matters deteriorated further, and it became something of an open secret that Professor Milberg and his colleagues might consider a deal to take Rover off their hands.

The problem was that no other major carmaker was interested, beyond the possibility that Land Rover and Range Rover might be made attractive if they could be hived off as a package. The Rover 75 was in full production at Cowley, Land Rovers and Range Rovers were streaming out of Solihull, and the Honda-related Rovers were likewise coming off the lines at Longbridge.

In the midst of this, MG still languished as a one-model brand, and the all-new Mini was still some way from production launch at the expanded and refitted Longbridge works. To be fair, BMW had merged Mini and MG in terms of sales and engineering, and was considering some future concepts.

Entrepreneur Jon Moulton of Alchemy Partners told the author how he saw, while on a skiing holiday in late 1999, the excellent range of 4x4 models BMW already had in its stable. Thinking that the company did not really need Land Rover technology, he put it to BMW that they could sell off Land Rover to defray some of their other

OPPOSITE: The face-lifted MG saloons along with the MG XPower SV, photographed in 2004. *Left to right*, behind the SV, are the MG ZT, MG ZS, MG ZR, and MG *TF*. *Author archive*

ABOVE: All seemed well at Rover Group on the MG side in 1999, which saw among other offerings a 2000 Model Year face-lift (including the option of semiautomatic transmission) and this 75th Anniversary Limited Edition. *Author archive*

MIDDLE: A face-lift idea from March 2000 for the MG*F* by Tony Hunter. The design team was led by David Saddington. *Tony Hunter*

RIGHT: It would be a mistake to imagine that no work was done on a new MG sports car under BMW. Hunter was one of the leading British designers working on MG and Mini proposals in this period. This is one of his beguiling sketches from 2000, drawing on the history of MG's earlier days of breaking records. *Tony Hunter*

RIGHT: Conceived post-BMW, the MGF Trophy 160 was essentially a more powerful version of the MGF VVC, with stiffened, lowered suspension. At the same time, it retained the Hydragas system of the normal MGF, albeit with competition-based spring and damper rates, and a ride height lowered to 20mm (.79 inches). The front brakes were AP Racing units with 8-inch (304-mm) ventilated discs and red-finished, MG branded aluminum calipers. The car's vivid color in this press shot is Trophy Yellow. *Author archive*

hemorrhaging costs. This argument was enough to convince company leadership, and before long, BMW let it be known that it was planning to sell off the Rover and MG parts of the business—and that the Land Rover and Range Rover side would be going to the Ford Motor Company.

Before long, two rival suitors emerged. One, Moulton's Alchemy Partners, sought to preserve the Rover 75 under license and embark on a program to focus on the MG brand as an underused asset; they would also consider a new joint venture with Lotus to develop a family of low-volume sports cars with composite-material bodies. Moulton had a strong track record, and few doubted he could make his plan work, but the less palatable aspect of his plan was the bonfire of jobs it would have involved, particularly at Longbridge.

The other plan to take on Rover Group came from a new alliance between John Towers, Rover Group engineering director Nick Stephenson, and two more members who branded themselves as the "Phoenix Consortium," perhaps with a nod to the sports car–based projects that had led to the MG RV8 and MG*F*. As time passed, these men would become known, less reverently, as the "Phoenix Four."

At first, notwithstanding the understandable opposition from the unions representing the Longbridge workforce, the Alchemy plan appeared to be the more viable, not least because they were financially well grounded, while there appeared to be some doubts about how the rival Phoenix bidders would secure the funds necessary for their bolder plan to retain volume car making at Longbridge. A great deal of negotiation went on behind the scenes, and it was evident that the government of the time, both Prime Minister Tony Blair and his trade and industry secretary, Stephen Byers, favored a solution that preserved jobs, even if they were not prepared to weigh in with further government grants.

BELOW: The MG*F* SE, with its unique Wedgewood Blue paintwork, was announced in July 2000. Sold under the new regime, it had clearly been first conceived as a last gasp of the old order. *Author archive*

TOP: In January 2001, the newly formed MG Rover Group confirmed that it was working on a range of X-coded projects derived from the existing Rover range. Shown here is the X10, clearly based on the Rover 75, while the X20 and X30 were based on the Rover 45 and 25, respectively. *Author archive*

ABOVE: In addition to the higher-performance Trophy 160, MG Rover also announced a version with a smaller 1.6-liter K-Series engine, the MG*F*1.6. *Author archive*

In March 2000, BMW finally announced that it would sell the Rover Group to the Phoenix Consortium. BMW retained the Mini name, along with Riley and Triumph, though neither of the latter has returned to life in the quarter century since the deal closed. They disposed of the MG brand, though, along with other, older names in the cupboard such as Austin, Morris, Wolseley, and Vanden Plas. Land Rover was sold off to Ford, where it joined Jaguar in the Premier Automotive Group that Ford ran at the time. As a part of the deal with Ford, BMW retained a lien on the Rover name: They only licensed its use to the Phoenix Consortium, by way of ensuring the future integrity of the Land Rover and Range Rover brands. Eventually, Jaguar Land Rover would be sold off to India's Tata Motors in 2007.

Now that the deal had been completed, Rover and MG were both about to enter a new chapter in their history.

ANOTHER NEW PATH: FORMING MG ROVER GROUP

The rump of the Rover Group business, divested of the Land Rover and Mini elements, was acquired for a symbolic ten-pound note by Phoenix Consortium. As the now rebranded MG Rover Group, the company contracted to

LEFT: An independent business, Powertrain Products (PTP), often worked alongside Rover. Neil Butler, one of the PTP team, recalls: "We showed a modified Rover 200 with 1.8-liter K-Series offering 165 horsepower, with close ratio gearbox updated suspension and an MG badge in the front, the week that Phoenix Venture Holdings took over Rover Group. MG Rover's Nick Stephenson said 'You've just saved me years of development. . . .'" *Neil Butler*

BELOW: The eventual production model which developed from the inspiration of the PTP car above was the MG ZR (via MG Rover's X30 concept). The model shown is the top of the range MG ZR160, in striking Trophy Yellow. *Author archive*

the former Austin factory at Longbridge, while BMW kept the plant at Cowley, where it would go on to build the new Mini, its new small hatchback. A major operation was undertaken to transplant the Rover 75 from Cowley and prepare to install the forthcoming Mini lines in its place. As it turned out, this was not so different from Reitzle's plan.

The MG brand was to be reinvigorated by the MG Rover Group. The corporate entity, with MG now part of its official name, embarked on updating the MG*F*. First it would make a faster, more firmly sprung MGF Trophy (see photo on page 201) and a lower-priced MGF 1.6 variant. Next, in 2001 it changed the suspension to a new bespoke coil-spring system in place of the earlier fluid Hydragas suspension to create the MG *TF* sports car, bringing back a name from MG's glorious past. The changes to this last model were extensive, an unusually high number for any car, and included replacing an entire suspension system and at the same time making many body alterations.

ABOVE: Also unveiled at the 2001 Geneva Salon was a station wagon version of the MG ZT, called the MG ZT-T. The second T stands for "Tourer." *Author archive*

RIGHT: The MG range photographed in 2001 at Longbridge. *From left to right*: the MG ZT-T, MG ZT, MG ZS, MG ZR, and MG*F*. *Author archive*

MG Rover set about applying the MG badge to the kind of sportier, BMW-baiting Rover derivatives that Munich had frowned upon. MG-badged, sportier-tune versions of the Rover 25, 45, and 75 arrived as the MG ZR, ZS, and ZT. Also, largely as a legacy of the BMW-era engineering program, an estate tourer (station wagon) was offered for both the Rover 75 and MG ZT—the latter became the MG ZT-T. Key to most of this work were a new design director, internationally renowned designer Peter Stevens, and a new engineering director, Rob Oldaker, a former senior Rover engineer who was persuaded to return to his old haunts after a spell at Bentley Motors.

One of the more surprising moves was the introduction of a commercial version of the MG ZR, badged as the MG Express van. Observers were further amazed that MG Rover found the time and the funds to reengineer the 75/ZT models to accommodate the Ford Mustang 4.6-liter V-8 and rear-wheel drive—even in tourer form. In the midst of this frantically busy period, the MG Rover board also bought the assets associated with the short-lived Qvale Mangusta, with a view to creating a kind of MG supercar. Somewhere in this plan, at least initially, lay thoughts of a return to U.S. sales along with the MG *TF*, possibly with a bought-in powertrain.

BMW had provided a dowry when it disposed of Rover Group, certainly, but this cash pile was being burned through all too swiftly. The new management at MG Rover Group began an at times desperate series of negotiations to form a meaningful relationship with another large automotive business to shore up its finances. It was no surprise when, as BMW had found in the previous decade, few of the obvious big names wanted to play, other than perhaps as component suppliers or customers; typically the powertrain found the readiest takers.

The most distinctive volume project of note at this time was the launch of a new small hatchback built by Tata of India, badged as the City Rover, appearing in late 2003. There were talks of a deal with a Chinese carmaker, China Brilliance, which faltered for various reasons, and of a relationship with Proton of Malaysia. A negotiation that did have legs was one with the Shanghai Automotive Industry Corporation (SAIC), headquartered in China and already that country's most successful carmaker. This company had built highly lucrative joint ventures with GM and VW.

RACING AGAIN

In both the Alchemy and Phoenix bids for Rover Group, there had been mention of rebuilding the MG brand and reentering the motorsports arena in a big way. The 24 Hours of Le Mans was brought up as a way to help each bidder make an impression and underscore their broader objectives. With MG Rover in place, the next step on this racing path was the creation of a new subsidiary of MG Rover parent Phoenix Venture Holdings, to be called MG Sport & Racing Limited. With this came a new sub-brand, called MG XPower. Both subsidiaries were unveiled at a press conference on April 26, 2001.

These plans were extremely bold and ambitious. While most expected continued support for MG *TF* racing, the new plans included rallying with a version of the MG ZR, an entry in the British Touring Car race series with the MG ZS, and, at the pinnacle of Sport & Racing's efforts, a new Le Mans LMP500 race car, codeveloped with Lola Cars, and a new two-liter, MG-branded MG XP-20 race engine developed by AER. In addition to these, an artist's impression of a proposed super MG*F* appeared on May 10, 2001. Perhaps it was the spiritual successor of the aborted MG*F* Super Sports discussed above, now dubbed the MG*F* Extreme.

The renewed EX Register sequence established in the BMW era was also maintained. At the time of writing, the sequence, restarted at EX257, had extended as far as EX265, as shown in the table below.

The first iteration of MG Rover's thorough face-lift and reworking of the Qvale Mangusta was the X80 concept, designed by Peter Stevens. It was unveiled to the public at the Frankfurt Motor Show on September 11, 2001, an event sadly overshadowed by the terrorist attacks in New York and Washington, D.C. *Author archive*

EX Vehicles of the Early 2000s

E CODE	PROJECT	DATES
EX257	MG Lola Le Mans sports car with MG XP-20 inline 4-cylinder turbocharged and intercooled 2.0-liter engine	April 2001
EX258	MG ZR Super 1600 rally car with K-Series 1.6-liter engine	April 2001
EX259	2001 MG ZS touring car with KV6 2.0-liter V-6 engine *[MG ZS touring car with KV6 2.0-liter V-6 engine, under Lola Engineering design direction but built by West Surrey Racing and with engine prepared by AER]*	April 2001
EX260	2002 MG ZS touring car with KV6 2.0-liter V-6 engine *[2002 version of MG ZS touring car with KV6 2.0-liter V-6 engine, built by West Surrey Racing and with engine prepared by AER]*	February 2002
EX261	2003 MG ZS touring car with KV6 2.0-liter V-6 engine	January 2003
EX262	2004 MG ZS touring car with K4 2.0-liter K4-cylinder engine	March 2004
EX263	MG ZR Group N 1.4 rally car (planned)	June 2005
EX264	MG Lola Le Mans sports car with Judd V-8 engine	December 2004
EX265	RLR Motorsport—MG Lola EX265—Judd	January 2008–2012

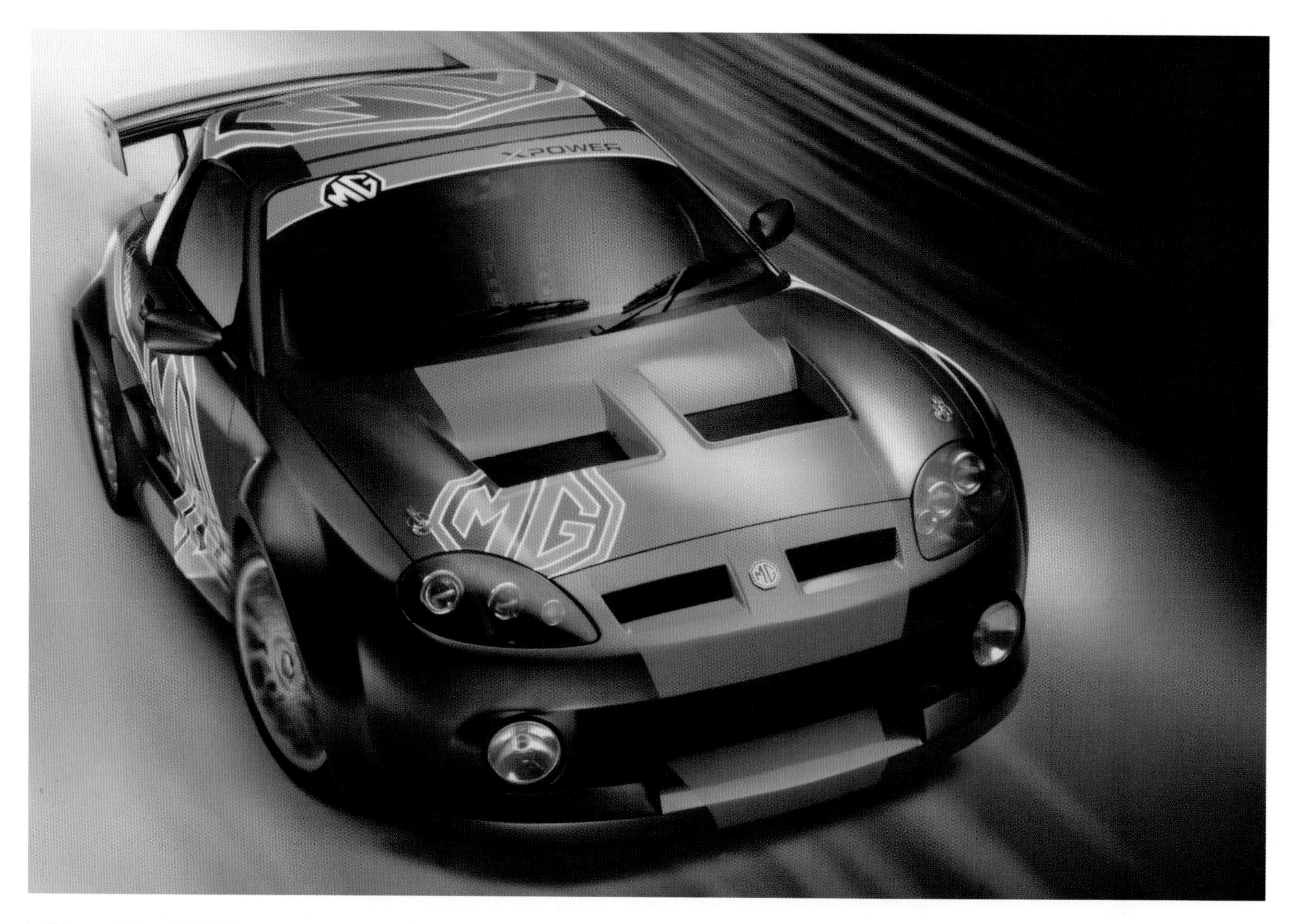

The exciting MG*F* XPower 500 was a reengineered MG*F* with a two-liter AER engine offering 500 horsepower and a mildly restyled front end that hinted at innovations to come. This car was later renamed as the MG *TF* XPower 500. *Author archive*

The Le Mans efforts took place in both 2001 and 2002; the second year also saw an entry by a private team (Knighthawk Racing) alongside the factory team effort. Both years were exciting, drawing enormous patriotic support from flag-waving Britons at the event. Neither event brought the hoped-for victories, though, and financial constraints prevented a third Le Mans attempt.

ONE-AND-A-HALF MILLION MGs

In the midst of the MG Rover renaissance there were a number of great reasons for celebration. One of these with an MG theme was a one-off MG *TF* 160, finished in a bespoke Jubilee Gold Supertallic paint finish and specially embroidered seats. The idea came from MG Rover's Corporate Affairs department, in particular Kevin Jones of that team, as a way to celebrate both the Golden Jubilee of the reign of Her Majesty Queen Elizabeth II and the landmark (in production terms) of a total of 1.5 million MG cars since records had begun.

Announced on April 16, 2002, MG again made reference to the official 1924 "origins" of MG explained in chapter 1. It was a good way to elevate the prestige of the MG name and makes an interesting comparison to the millionth MG made at Abingdon of 1976 and other statistics, given below.

While work was progressing in fits and starts on a new mainstream model range, the next anniversary MG *TF*—the U.K.-only 80th Anniversary LE—was being readied for launch on January 27, 2004. MG Rover had settled on May 1924 as the date of origin for MG, and so the new limited edition, with just five hundred offered, was made available in a choice of three colors; two of them new (Pearl Black and Goodwood Green), as well as the existing Starlight Silver. The LE also offered a choice of 135, 120 Stepspeed (CVT transmission), or 160 power units. The silver and black cars were teamed with burgundy soft tops and Ash Gray interior, as well as Grenadine Alcantara®

ABOVE: At the launch of the MG ZR, ZS, and ZT ranges at St. David's, Wales, in July 2001, MG Rover displayed this "extreme" version of the ZT, the ZT X500, with a 500-horsepower Ford V-8, shown here in XPower green-and-gray livery. *Roger Parker*

LEFT: One of the more remarkable projects taken on by the new MG Rover team was reengineering the usually front-wheel-drive MG ZT platform to take a Ford Mustang V-8 driving through the rear wheels. It was launched as the MG ZT 260. *Author archive*

and black leather seats, while the green ones had tan leather inside and a matching tan soft top. Much was made of the fact that a single specimen of this new LE was auctioned for £275,000 at a charity ball.

The same month, January 2004, also saw face-lifts across the MG ZR, ZS, and ZT ranges, as well as their Rover contemporaries. These showcased sharper styling, developed by Peter Stevens and his team. They offered a neat, effective look, though one that could not disguise that they were smaller platforms, largely based on Honda technology, that were already a decade old; even the newer 75/ZT was heading for its sixth year in production. MG Rover was still working on a new RDX60 model and derivatives of the 75 platform, which it hoped to entice SAIC. In fact, negotiations with SAIC still looked promising in the summer of 2004, with MG Rover revealing in June that joint studies had commenced.

In the same year, show cars were created under Peter Stevens's direction, including a fixed head coupe version of the MG *TF* and a stunning two-door coupe version of the Rover 75. The MG *TF*'s design was predicated on a 2.5-liter KV6, although the prototype—which the author drove—made do with a K-Series four. Both the MG *TF* and the Rover 75 were one-off, hand-built concepts intended to demonstrate the creative thinking inside MG Rover, but none had serious production development work behind them. They were exhibited alongside a specially trimmed version of the MG XPower SV in the autumn of 2004. A proposed MG Midget concept, also described at the time, was never turned into a full-size model as had originally been intended.

By the spring of 2005, it was five years after the BMW divorce and thirty years since British Leyland's initial foray into public ownership. The cash taps at Longbridge were running dry. MG Rover had embarked on Project Drive, a costs-down, "de-contenting exercise" of a kind that is so often a surefire indication of icebergs ahead.

MG Rover was limping on, desperately trying to broker a hard-cash deal with SAIC, and had gone through such acts of desperation as selling off its remaining real estate to a major U.K. property developer and leasing back the land on which its Longbridge factory sat. For SAIC, it had clearly become a waiting game until the body went cold.

THE END OF MG ROVER

MG Rover went into administration on April 8, 2005, when the deal with the SAIC collapsed and the company did not have sufficient cash to continue trading. SAIC had been waiting on the wings for this moment. No doubt as concerned in its own way as the British government was about MG Rover's growing precariousness, the firm was accountable to its own board of

TOP LEFT: The MG Rover Group did not take long to develop and launch a significantly restyled and reengineered sports car, the new MG *TF*, launched on January 8, 2002, and developed from the MGF. The suspension now used steel coil springs in place of Hydragas. *Author archive*

LEFT: The U.K. wasn't the only country to see special editions of the MG *TF*. This is MG *TF* 160 Vintage Racing Limited Edition, created for the French market. *Author archive*

management and the Chinese government, its principal shareholder. From the National Audit Office report, "The Closure of MG Rover" (NAO, 2006), we learn that "on April 15, 2005, the [Department of Trade and Industry] received a letter from SAIC in which it said explicitly that it was not willing to acquire either the whole or part of the MG Rover as a going concern."

With MG Rover's principal potential business partner unwilling to buy the assets, the business was placed in administration, with a number of potential bidders declaring their interest in some or all of the business. There is insufficient space to do justice to this process, which involved the various interested would-be buyers trying to convince PricewaterhouseCoopers of the viability of their various offers. From an MG enthusiast's perspective, some of the offers and plans listed seem exciting—for example, the inventive "Project Kimber," which would have delivered a much-modified MG *TF* in addition to a reengineered Smart Roadster, the latter rebadged as a new MG Midget—but ultimately the administrator saw his role as securing the best outcome for the creditors, a path that so often precludes more adventurous opportunities.

The MG-Lola EX 257 Le Mans car was part of MG Rover's bold plan to enter motorsports. Later iterations became the EX 264 and EX 265. *Author archive*

TOP LEFT: The MG EX 258 MG ZR rally car in factory MG XPower livery. *Author archive*

RIGHT AND BELOW LEFT: The third in the new era of EX Register codes was the EX 259, a circuit-racing version of the MG ZS in four-door saloon form. A later evolution of this model brought the EX 260, EX 261, and EX 262. *Author archive*

XPOWER
MG
AA www.theAA.com
MSA British Touring Car Championship
2
MG
XPOWER
BFGoodrich
BFGoodrich
WSR

ABOVE: The production version of the MG XPower SV evolved from the design of the earlier X80, itself developed from the platform of the Qvale Mangusta. This is the MG XPower SV-R, a more potent version of the SV. *Author archive*

RIGHT: Peter Stevens and Nick Stephenson were behind this project, whose goal was to create the world's fastest station wagon. Dubbed the X15, this car was developed with the famous So-Cal Speed Shop and used a Ford-based V-8. *Peter Stevens*

THE NANJING AUTOMOTIVE STORY

In this curious interregnum, some unexpected developments led to the MG Rover legacy being split into two parallel threads. On July 27, 2005, another Chinese automotive company, Nanjing Automotive Corporation (NAC), stepped in and quietly purchased most of the production equipment that was now sitting idle at Longbridge, thereafter lifting and shifting much of it over the early part of 2006 to a new plant at Pukuo, Nanjing Province, along with, in a strange twist, the intellectual rights to the MG badge.

What may have been most interesting from an international, especially U.K., perspective was the decision by NAC to restart production of the MG *TF* sports car. Discussions were held with U.K. company Stadco, effectively the successors to Mayflower, the bodyshell producers who had codeveloped with Rover Group the original MG*F* tooling in 1993. A deal was struck, and all the new MG *TF* bodies were set to be manufactured in China; for U.K. sales, the painted and partially trimmed units were to be shipped all the way to Longbridge for final assembly on a refurbished line there.

In the meantime, however, there was a short-lived proposal to create a new factory in Oklahoma, with the idea of flying in parts via the nearby commercial airport and building a coupe version of the MG *TF*, loosely modeled on the one-off created by Peter Stevens. There was a brief puff of international excitement at this idea, following a big local launch in the United States involving local civic dignitaries and the NAC management on July 12, 2006. Five days later, at a U.K. press conference, the NAC chairman poured cold water on the project, saying that it was no more than "an idea" being considered. Needless

to say, the project never progressed any further, and the MG *TF*, in either roadster or this new coupe form, never made it to U.S. showrooms.

While NAC continued firming up its production plans with MG, SAIC had not been idle, having taken the designs that it had acquired and turning the Rover 75 into the basis of a new model. This car, the Roewe 750, was based to some extent on previous face-lift plans that had been in the pipeline at MG Rover. The Roewe name was a brand-new, self-owned marque that came about at least partially because SAIC was unable to persuade the custodians of the Rover name to allow them to use it.

The Roewe 750 closely resembled the Rover 75, but it sported a much-altered fascia, a slightly longer wheelbase, and a new version of the Rover KV6 2.5-liter petrol engine, built at SAIC's Baoshan engine plant. SAIC was working in the U.K. with a dedicated facility, staffed in part by former Rover engineers and designers. One of their key tasks was the development of a new model, the Roewe 550, whose relevance to the MG story will become apparent later.

THE RETURN OF THE MG *TF*

Back in the world of NAC and MG, the formal unveiling of the new MG *TF* line at Longbridge was held on May 29, 2007, in tandem with a start of low-volume assembly in Pukuo for sales in China—not at the time a market noted for its appetite for open-topped sports cars.

Later in 2007, production of the MG ZT began again in China, now under a new model name, the MG7. This car was launched on September 28, 2007. To all intents and purposes, the MG ZT and MG7 were almost identical, beyond the badging and some minor trim changes, and as the ZT may have been one of the best MG saloons built for many years, the new MG7, too, won some fresh praise and custom, albeit as a vehicle that was now only sold in China. The fact that it also resembled, but was quite separate from, SAIC's Roewe 750 was a matter of some comment: Both the MG7 and Roewe 750 offered versions of the KV6 engine design, but each was separately engineered and manufactured.

The Chinese marketing team decided that MG—pronounced *minju* in China—needed a new strapline. A canny marketer, whose name has been lost amidst the subsequent laughter, decided that the MG letters stood for "Modern Gentlemen" rather than "Morris Garages." This strange idea had a thankfully short life.

ABOVE: The face-lifted MG ZS was perhaps the most surprisingly successful overhaul of the Z range of MG cars. The 2.5-liter V-6 version was a great sporting car to drive, with an engine to rival that of an Alfa Romeo. A wistful example of what was lost with the sad end of the MG Rover story. *Author archive*

BELOW: The Golden Jubilee of HM Queen Elizabeth II took place in 2002, and MG Rover decided to build a one-off MG *TF* 160, finished in a new Monogram Supertallic paint finish, coincidentally and appropriately dubbed Jubilee Gold. Note the use of Old Number One again as part of the marketing effort. *Author archive*

RIGHT: The one-off MG GT was conceived and built by Peter Stevens in a crash program with Dove, a small specialist prototype shop in Norfolk. Although the marketing write-up implied it was expected to use MG Rover's 2.5-liter V-6, the prototype made do with the usual four-cylinder powertrain—as noted by the author when he drove it. *Author archive*

At the November 2007 Guanghzhou Motor Show, NAC followed on with an MG-badged version of another MG Rover creation, the Rover Streetwise. This was an "urban styled," higher-riding version of the Rover 25, which in China became thc MG3SW, featured alongside the returning MG *TF* CVT. Sales were again limited for both models to the domestic market in China.

It had already become evident that the Chinese government, the ultimate governing influence for both NAC and the much larger SAIC, sought to bring the divergent MG and Roewe efforts under one roof. NAC had built its modest bridgehead to Europe, through the Longbridge plant. As the lead national automotive manufacturer, there were clear expectations that SAIC would not only build its own expertise, much learned from its joint ventures, but also embark on an ambitious export program over the coming decade.

Perhaps it was inevitable, therefore, when on December 26, 2007, it was announced that

MIDDLE LEFT: The MG7 was little more than a mildly modified MG ZT, built and sold in China between 2007 and 2013. At the same time, SAIC built a Roewe 750, which looked very similar to the MG7 but had been engineered separately. *Author archive*

LEFT: A stillborn MG Midget, part of the X120 project, one of the many tragic consequences of the eventual collapse of MG Rover. In the author's view, this could have been a beguiling small MG sports car, and even U.S. sales might have proved feasible with a proposed engine transplant. *Author archive*

the automotive interests of NAC and SAIC would be merged, with the larger company obviously the dominant partner. The chairmen of SAIC and Yuejin—Hu Maoyuan and Wang Haoliang, respectively—signed the cooperation agreement in the People's Congress Hall in Beijing in the presence of the Chinese vice prime minister and various civic dignitaries from Shanghai, Jiangsu Province, and Nanjing's municipal government.

This move initiated the present, nearly runaway success of a revitalized MG. Immediately after the signing ceremony, SAIC, a Global Fortune 500 company and the most prodigious carmaker in China, took full control and ownership of an internationally recognized name. In the background, SAIC had as noted retained the services of a jointly owned consultancy, originally a joint venture with U.K. engineering consultancy Ricardo, and based in a dedicated unit not very far from the MG Rover HQ. The SAIC team soon moved in to Longbridge, and a fresh round of design and engineering work began. That is the story for our final chapter of the MG centenary story.

ABOVE: Testing of the Nanjing MG *TF* took place in China. This test car is photographed in Turpan (Tulufan) in China's autonomous Xinjiang Province, home to desert and the Flaming Mountains. In testing terms, it is roughly equivalent to the Nevada desert in the United States. The MG *TF* was relaunched in the Chinese market but found few takers. *Author archive*

BELOW: This concept sketch depicts an abandoned proposal for a new MG *TF* coupe, intended for assembly in a new U.S. plant in Ardmore, Oklahoma. Despite a grand agreement ceremony, the project went nowhere. *Author archive*

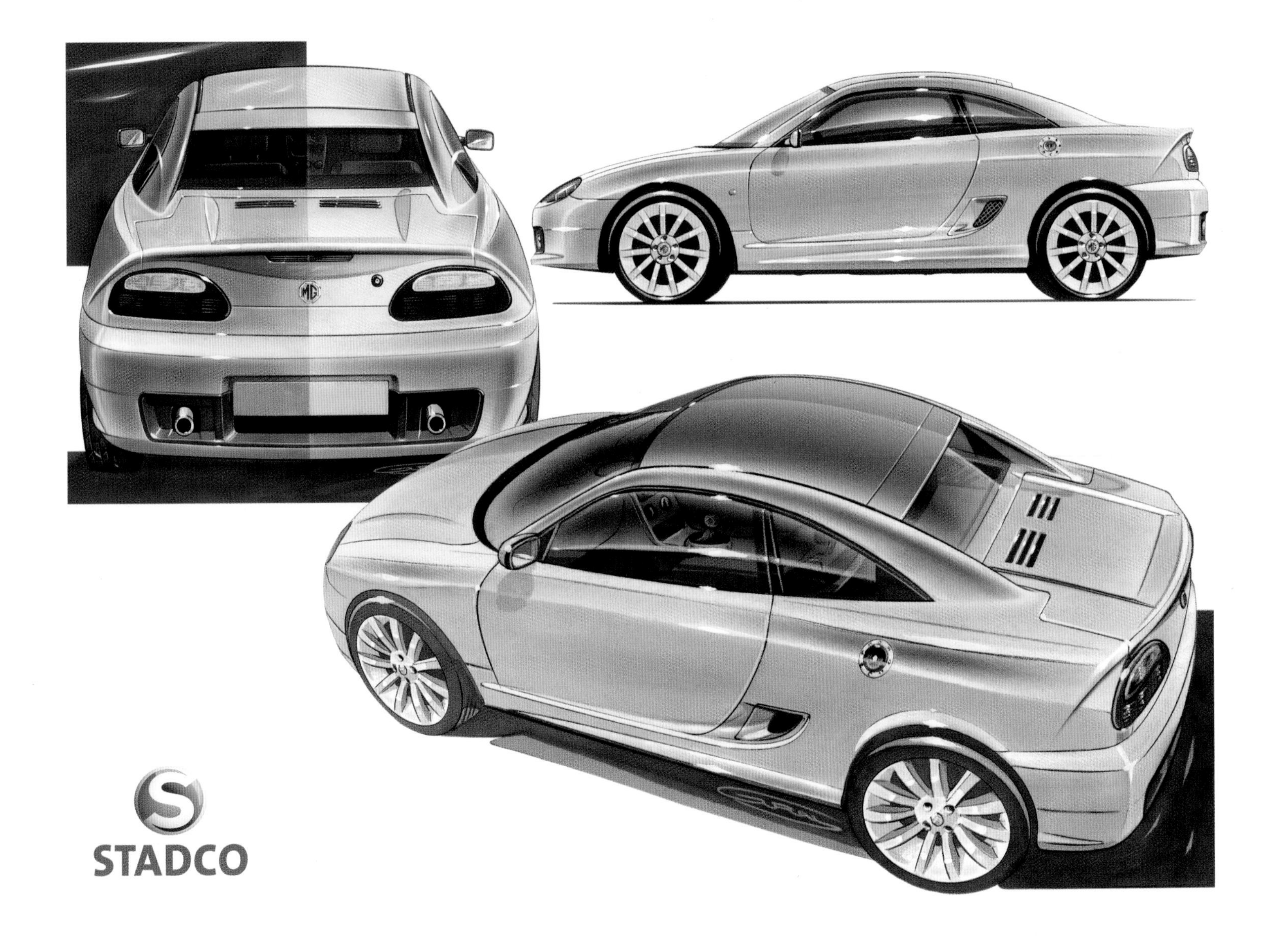

MG
MG6 XPOWER

10

Renaissance: Toward a Glorious Centenary

THE NEW GENERATION OF MGs

The integration of the NAC and SAIC interests did not take long to be realized. The SAIC facility at Leamington Spa was closed and its personnel moved across to Longbridge, working alongside their NAC counterparts. The first key task was the development of new MG models that would sell well not only in China, where the public appetite for cars with a Western flavor was growing fast, but also in export markets, where SAIC (thanks to its own shipping line) had the clout and contacts to forge international success.

NAC MG U.K. changed its name to MG Motor U.K. on January 8, 2009, and the main SAIC Motor U.K. Technical Centre (SMTC U.K.) moved over to Longbridge from Leamington Spa to March 3, 2009, as part of an investment of £2.7M. A new studio at Longbridge opened later in the year, led by British designer Tony Williams-Kenny. The first fruit of this unified team was the MG6, a model closely based on the platform of the Roewe 550 and which had been largely coengineered at Leamington Spa and SAIC's Anting complex near Shanghai.

The MG6 made its much-anticipated show debut at the Guangzhou Auto Show on November 23, 2009. There were now three versions of the latest MG *TF* on sale in the U.K.: the *TF* 135, *TF* LE500, and *TF* 85th LE. The limited-edition car was yet

OPPOSITE: November 2009 saw the unveiling of the MG6, SAIC's first self-developed MG, designed with the support of the U.K. operation. This is the first U.K.-built MG6 at Longbridge on April 15, 2011. Production of this model at the Birmingham plant effectively replaced that of the MG *TF*. MG Motor U.K. also sold a diesel version for a time. *Author archive*

another anniversary celebration, drawing on the 1924 badge registration. The badge itself was revamped, with its theme reverted from brown and cream to black and white. In China and some of the other new export markets where MG was emerging (in some cases, for the first time ever), there was once again reference to the "Morris Garages" brand origins, though in markets like the U.K., where the MG name was already well known, it was not deemed necessary to revisit that story.

The MG6 became key to China's bold plans for exports, using the benefit of a familiar British brand name. It was hardly a coincidence that MG badges were also applied to unrelated Roewe and Baojung models for markets such as South America, Africa, and the Middle East. With a similar strategy, SAIC began building Buick-badged SUVs for North America—but no MGs were involved in that particular plan. Like its neighbors in Japan and Korea, SAIC has built its knowledge and capacity carefully but unremittingly. MG models have been sold in some markets where there was perhaps less of a need to stick to traditional brand tropes. This meant that there were models badged variously as MG550, MG5, MG750, and MG RX8, often mixing and matching MG and Roewe models but with no more to distinguish them than the badges. It is an echo, perhaps, of the days of BMC, half a century earlier.

By this stage, the Longbridge plant was building some of the last examples of the MG

TOP: Preproduction of the new Longbridge build of the MG *TF* in 2007. *Author archive*

ABOVE: The first of the new MG *TF* LE 500 models is driven out of the Longbridge Assembly Building to fanfare, fireworks, and balloons. *Author archive*

RIGHT: The final MG *TF* (the white car) comes off the line at Longbridge, accompanied for the occasion by the first MG*F*. *Author archive*

LEFT: In China, the former Rover Streetwise, developed in the MG Rover era, proved popular in China with MG branding as the MG3 SW. It was launched in 2008 but never exported. *Author archive*

BELOW: Renewing connections to a beloved name from MG history, the MG6 Magnette four-door sedan arrived in May 2011. The rear bodywork was partly derived from that of an earlier MG550 Sedan that had been sold in South America. With the arrival of the Magnette, the five-door MG6 became known as the MG6 GT. *Author archive*

TF, as the Birmingham factory was readied to produce the first U.K.-assembled MG6 models. The initial five-door GT hatchback would be followed by a four-door MG Magnette, both with a 1,796 cc TCI Tech petrol engine and later a new, bespoke 1.9-liter diesel engine. The engineering and design functions were linked to their international counterparts in China and a new concept for a small hatchback, code-named MG Zero and unveiled at the April 2010 Beijing Auto Show, previewed what became the all-new MG3. This was the basis for the second big push for MG exports.

Guy Jones, MG Motor U.K. sales and marketing director at the time, declared that MG Zero was "crucial to the development of the brand globally, as it gives clear direction for the future beyond the current products. We are all proud to see our British-designed vehicle wearing the MG badge creating such an impact in Beijing. With the MG *TF* back in production in Birmingham, sales increasing and the MG6 coming at the end of the year, the MG Zero has come at the perfect time to build further awareness and interest in the brand here in the U.K." As Jones was speaking, preparations were in hand for the opening of a new £5

MG Badged Roewes

TOP AND ABOVE: In the interest of exploiting the MG name as an export brand, the idea of selling MG-badged Roewe models took flight in a number of markets where perhaps the heritage of MG was less well known. In sedan terms, this resulted in such confections as the MG350, MG360, MG550, and MG750 (shown here, as sold in Egypt), while SUV offerings in some markets included the MG RX8 and MG RX9. *Author archive*

million ($8 million) design center at Longbridge, built on the site of the former MG Rover Sales & Marketing building.

The MG6 production line at Longridge was officially launched on June 26, 2011, during a visit on that day by Chinese premier Wen Jiabao. It was declared the first all-new MG model to be launched in the U.K. in sixteen years, the last to fit that definition having been the MG*F*. As expected, the MG Zero concept begat the production MG3, launched in China in 2011 and assembled initially both there (at the former NAC plant at Pukuo) and, in mildly face-lifted form, in Longbridge after 2013. Some markets even saw an offshoot, the slightly higher-riding MG3 Xross, aimed at the customers who had bought the NAC-era MG3 SW described above.

Before long came news of a joint venture between SAIC and a major industrial partner in Thailand, which opened in 2014 and expanded into a second plant five years after that. Similar to the MG Zero before it, another British-designed concept, the MG Concept 5, had appeared at the following year's Auto Shanghai (the Beijing and Shanghai annual shows generally alternate); it was a preview of the new MG5 midrange hatchback, a roughly VW Golf–sized model with McPherson strut front suspension and a simple twist-beam setup at the rear. Although widely exported and built in Thailand as well as China, this first MG5 (known in some markets as the MG GT) was never imported to the U.K., despite initial suggestions that it might have been marketed there.

By this time, MG design was being led internationally by Tony Williams-Kenny, who shuttled between Shanghai and the U.K., while his U.K. studio was overseen by his deputy, Martin Uhlarick. Designer Carl Gotham was another key member of the team; he was partly responsible for the Concept 5 study.

As a follow-on to the MG Zero and MG Concept 5 show cars, which had provided strong clues to imminent production cars, SAIC sought a more radical centerpiece for the MG display at the April 2012 show, now back at Beijing. The MG Icon was a radical blend of classic MGB design cues, a modern cross between a small SUV and a sporty hatchback. The work was largely overseen by a small team led by Stephen Harper, a design consultant whose early career had been spent at Longbridge in the Austin Morris era. The design was certainly one of the talking points of the Beijing show, although it was evident that the concept was not linked to any definite production plans. In response to the inevitable questions about a new MG sports car, it was

explained that such a car remained some years in the future, as was any prospect of a return to the U.S. market. A few weeks before the MG Icon was shown at Beijing, the *Birmingham Mail* reported that SAIC had already invested a total of £450 million in MG Motor.

The MG3 received a moderate face-lift in 2013 and, in June, the U.K. Design Studio benefited from an additional £1.5 million ($2.4 million) investment, opening its doors to the media, where we learned that it was now the fifth largest design studio in the U.K. The MG3 was now being marketed in the U.K. with a wide range of "personalization" decal treatments aimed at drawing in a younger customer base. It has continued to sell quite strongly and received a suite of more comprehensive revisions in 2018; as of 2023, it was still on sale in the U.K. as one of the lowest-priced petrol-engined cars on the market. In some markets it is offered with an automatic transmission in lieu of the standard manual gearbox, but its future seems finite as the world moves toward hybrid and fully electric automotive powertrains.

TOP: A bold new concept for a small family hatchback, the MG Zero previewed what would become an all-new MG3. *Author archive*

ABOVE: The MG Concept 5 previewed the first-generation MG5, which was sold mostly in China and emerging markets like South America. It never came to Europe. *SAIC Design and Author archive*

ABOVE: The MG Icon concept of 2012 was intended to be a blend of classic MG sports car and modern sports SUV design themes. Those unusual vertical strip lamps take their visual cue from the MGB over-riders. Despite much interest and a flurry of speculative rumors, the Icon did not translate into a new production MG. *Author archive*

RIGHT: The MG3 received a second face-lift in 2018, adding a touch of chrome and the then-current MG grille style. It was still available in this form five years later. It's proved to be a popular first car for many new drivers. *Author archive*

LEFT: The 2014 MG CS Concept previewed SAIC's first all-new family-size SUV. Note the narrow headlamps and slim grille aperture, as well as a large MG badge. *Author archive*

MIDDLE: The MG GS of 2016 was the production model that the CS was intended to preview. Although there were plans to build this model at the U.K. Longbridge plant, the idea was dropped in favor of full imports, and the U.K. production lines fell silent again. *Author archive*

BOTTOM: 2014 called for another official MG anniversary—MG 90—which in the U.K. was marked by this special-edition version of the MG6 GT. *Author archive*

Racing Still Improves the Breed

Motorsports have always been at the heart of MG, right from the first Morris Garages efforts back in 1923. As we have seen, throughout the Rover Group and MG Rover Group eras, various motorsports activities continued, so it was gratifying when MG Motor U.K. announced their support of a brace of specially developed MG6 race cars in the 2012 British Touring Car Championship (BTCC).

For the first season, the cars were run by MG KX Momentum, with drivers Jason Plato and Andy Neate. Many sceptics may have been surprised when Plato gained third place at season's end. For 2013, the cars were run by Triple Eight, with Sam Tordoff in his first year alongside Plato; they finished the season in sixth and third places, respectively.

In 2014 the team were understandably ecstatic at breaking Honda's four-year records at the Manufacturers' Championship. This year there was a third MG6 GT on the grid, in new livery. A road car SE, the MG6 BTCC Special Edition, was offered in a choice of blue or white with special trim but no performance changes.

Back on the tracks, a new driver lineup followed for 2015, with Andrew Gordon and Jack Goff. After a strong season, they took second place in the Manufacturers' Championship. MG Motor agreed to extend the contract in 2016 for a further three years, although the MG6 was also withdrawn from sale at that stage, and the last three seasons for the racing versions were comparatively disappointing after what had been achieved hitherto. The merits of a dedicated motorsports campaign were not lost on MG China, and there the MG XPower theme, in particular the bold graphics and a combination of Metallic Gray and Fluorescent Green Livery, was revisited, with a concept based on the face-lifted second-generation MG6 shown at the Shanghai Motor Show of April 2019.

Enjoying a new center of gravity in the Far East, MG entered the 2019 Touring Car Asia Championship series with the MG6 X-Power TCR under "Team MG X-Power."

LEFT: MG in the U.K. supported the MG6 in the British Touring Car Club racing series. When the sedan became the MG6 Magnette, the five-door became known as the MG6 GT. The MG6 GT continued racing even after the base car was withdrawn from the U.K. market. *Author archive*

BELOW: The MG XPower brand, complete with the same gray and bright green livery, was launched in China and used in racing in the TCR Asian race series. One minor change is the abandonment of the back-to-front letter R as the end of XPower. This was the MG6 TCR XPower for 2019, featuring a turbocharged DOHC four-cylinder of 1998 cc (bore x stroke: 88x82 mm), which offers 340 horsepower at 6,400 rpm and maximum torque 410 nm at 4,400 rpm. *Author archive*

For the 2013 Shanghai show, the MG offering was a CS concept, a medium-sized crossover that blended the emerging MG design language with a car that looked as though it was aimed squarely at Nissan's popular Qashqai, which had led an explosion of sales in the emerging tame SUV sector. Again the work of Williams-Kenny and his team, the CS effectively previewed the new MG GS production car, launched at Auto Shanghai in 2015 (two years after the CS) and in the U.K. in May 2016, at the reconstituted London Motor Show in Battersea.

There were more changes at Longbridge, where an emissions and component testing site for SAIC was now online. Assembly of the MG3 using semi-knocked-down (SKD) kits of painted and part-trimmed bodies ended quietly on July 5, 2016, when the last British-assembled MG, a black MG3, came down the line draped in a Union flag. In a muted irony, this ceremony took place two weeks after the U.K. public voted, on June 23, 2016, to leave the European Union.

There were still, as far as the public knew, plans to assemble the GS at Longbridge, like the MG *TF*, MG6, and MG3 before it, but on September 23, 2016, the Longbridge assembly lines were formally closed, bringing to an end 110 years of production there. There were still, for the time being, other design and engineering functions on the site, but MG3 and GS models had already begun arriving in the U.K. fully built, receiving predelivery inspection at Portbury Docks in Bristol before being moved through the distribution channels. It was

TOP: The MG GS was largely replaced by the later MG HS, though the model had a second lease on life in China in 2017 and then appeared in the Middle East for 2019 with a face-lift at the front and an upgraded interior. These changes brought it in line with the latest MG design language. *Author archive*

ABOVE: The second-generation petrol-engined MG5 Sedan, itself based on a Roewe model, formed the basis for the subsequent MG5 EV Estate sold in the U.K. and Europe. *Author archive*

RIGHT: The MG ZS was a turning point for the modern iteration of the MG marque. It was the first small MG SUV, with a new level of build quality and technology. *Author archive*

MIDDLE: The second-generation MG6, an attractive car denied sales in western Europe, seen here in its 2019 MG6 305 Trophy limited-edition guise. *Author archive*

BOTTOM: The MG e-Motion was a stunning concept for an electric two-door four-seat MG coupe, unveiled originally at the April 2017 Shanghai Motor Show. This is the first official photo of the vehicle, with SAIC Design's international design director Shao JingFeng. Although highly anticipated and praised by onlookers—and long expected to launch by around 2021—in the end it was a concept in need of a platform, and an adequate business case could not be made for it. One feature that did survive were the scissor doors, part of the later MG sports car concept. *Author archive*

LEFT: In China, an unusual publicity stunt to accompany the introduction of the ZS face-lift in 2019 was a high-speed run on the deck of a Chinese naval warship. The ZS was styled as a special XPower version of the new model. So far, no production equivalent has been seen. *Author archive*

a sad state of affairs, but in the background was the obvious fact that SAIC found it hard to justify servicing the costs associated with the lease obligations for the Longbridge site. A consequence of MG Rover's actions, by this time the lease was said to be around £49 million per annum. There could be no justification for the expense, when the company was using less and less of it, instead ramping up production in other factories overseas.

MG ZS AND HS: NEW MARKETS AND THE BIG SELLERS

Although the MG6 disappeared from the U.K. market, a second-generation version was unveiled at the end of 2016, its more modern design following the style of another new MG with much more of an international outlook, the MG ZS. The second-generation MG6 included a sportier version that brought back the Trophy badge, last used at MG Rover, and there was also a petrol electric hybrid (PHEV) version, the eMG6.

Maintaining the theme, a further face-lift of the MG6 appeared in 2021. This version, the MG6 Pro, was chiefly differentiated by the adoption of a new, more aggressive front-end styling, but again sales were not directed toward most of the mature markets outside China.

The much more internationally significant MG ZS was a new, small-to-medium-sized family crossover hatchback, previewed at the 2016 Guangzhou Auto Show and launched in production form at the April 2017 Shanghai show. Exhibited at the London Motor Show that summer, the marketing material and stand display proclaimed it as the MG XS, there being some concern within the U.K. team that a British audience might confuse it with the MG Rover–era MG ZS. Soon common sense prevailed and the model retained the same name in most markets, though a later version would appear in India as the MG Astor—of which more later.

Also at the 2017 Shanghai show was a truly stunning concept for an upmarket, all-electric, four-seater two-door MG coupe—the MG E-Motion concept—whose sleek lines drew admiring comparisons with the likes of Aston Martin and Audi. The electric powertrain was said to be good for a 0–62-mile-per-hour (0–62-kilometer-per-hour) time of less than 4.0 seconds and a range of more than 310 miles (500 kilometers). Although there were many

BELOW: The MG HS, the production car that followed the X-Motion concept. In early 2023 it became one of the best-selling cars in the U.K. Australia also has its two-liter four-wheel-drive version, which is not sold in Europe due to its emissions. *Author archive*

Design to the Fore

In the summer of 2019, three years after the end of MG assembly at Longbridge, came the news that SAIC was undertaking a global review of its operations. One of the casualties of this was the Birmingham SMTC. By this stage, far more of the fundamental engineering design and production work was being undertaken in China, and with no short-term resumption of local U.K. (or even European) assembly in view, the cuts were perhaps predictable if no less painful as a consequence. One brighter aspect was the news that SAIC would continue growing its still-new London design presence, the new SAIC Advanced Design London studio having been opened on September 19, 2018, under the management of Carl Gotham and Robert Lemmens.

LEFT: Although a dedicated studio had been established by SAIC at Longbridge in 2010 (with the subsequent retrenchment from that site), SAIC established an exciting new studio in the same central London HQ as MG Motor U.K. Under the leadership of Carl Gotham (seen here in front of a picture of the MG X-Motion concept), SAIC Advanced Design London works on future-thinking concepts for SAIC's self-owned marques, of which MG is one. *Author archive*

BELOW: A product largely of the London studio, which worked closely with counterparts in Anting, near Shanghai, the Cyberster concept for a modern all-electric MG sports car wowed the crowds at the April 2021 Shanghai Motor Show. It's seen here with Shao JingFeng at right. Several details, such as the headlamps, required changes for production. *Author archive*

TOP: The MG ZS EV in its first form looked very much like the normal gasoline-powered ZS. Externally the main clues were special alloy wheels, a bespoke color choice, and a charging point hidden in the radiator grille. For 2021, there was a styling refresh that dispensed with the standard gas ZS grille for the EV version. *Author archive*

ABOVE LEFT: The face-lifted version of the MG5 EV Estate car for Europe. *Author archive*

ABOVE RIGHT: Building on the XPower brand, which has been recreated in China, this is the 2021 MG6 XPower road car. The other image shows the car at the 2021 Shanghai Motor Show, with a display behind of XPower-branded accessories. *Author archive*

LEFT: SAIC has been pushing hard to capture the rapidly burgeoning Chinese youth market. The young girl on the poster is Yang Chaoyue, a well-known female influencer who was appointed as an MG Brand Ambassador in 2021. *Author archive*

positive noises about the E-Motion leading to a production car, it eventually fell victim to the simple fact that it was a lovely-looking concept in search of a nonexistant platform; arguably some design elements went into the later all-new MG7.

Following the by now established pattern, yet another MG concept appeared at the 2018 Beijing Auto Show. This was the MG X-Motion, again largely the work of the U.K. design team, which was now under the leadership of Carl Gotham after Tony Williams-Kenny had moved on to another role outside SAIC. (Martin Uhlarick had meanwhile moved on to become head of Tata's U.K. design studio in 2016.) In China, the MG and Roewe design functions were now overseen by Shao JingFeng, the new international design director whose previous role had been with the large and important SAIC-VW joint venture design.

The X-Motion was to many eyes one of the best SAIC-era concepts yet, and it closely informed the MG HS production car that reached markets later in 2018, superseding the GS in most countries. In contrast, a heavily face-lifted GS was sold for some time in China alongside the new model. Some markets, such as in Australia, would see a two-liter four-wheel-drive version of the HS; for most markets, including the U.K.,

TOP: The MG One was launched in China in the summer of 2021. To date it has not been exported. The color is called Bubble Orange. *Author archive*

ABOVE: Only meant for sale in China initially, the MG6 Pro featured a new front end similar to that of the MG One. *Author archive*

RIGHT: Sold only in China at the time of writing, the new-generation MG7, previewed at the 2022 Chengdu Motor Show, is the top sedan model in the MG range. Build quality is some of the finest ever seen on an MG of the modern era. *Author archive*

the same 1.5-liter SGE four-cylinder was the only powertrain on offer, with front-wheel drive and a choice of manual and automatic transmissions. A petrol-electric PHEV hybrid soon joined the range, soon to form part of MG's expansion into continental Europe. In 2023, the MG HS was face-lifted to meet evolving customer expectations.

With the inexorable march toward electrification for most of the global cars and commercial fleets in the run-up to 2030, electric designs have inevitably featured high on the Jing Feng SAIC (and MG) agenda. An EV version of the MG ZS appeared at the 2018 Guangzhou Auto Show, initially described as the MG EZS but later more usually called the MG ZS EV. The combination of the standard petrol ZS and the new ZS EV saw sales of MG rocket in many mature markets.

The primary work of the studio is to look toward the future, collaborating with other global SAIC studios on exciting, forward-looking concepts, for not only MG but also Roewe and the new R brand that SAIC added to its portfolio as an upmarket EV offshoot of Roewe. One of their more interesting projects, to MG fans at least, was surely their role in designing a concept for an all-new electric sports car, the MG Cyberster, which appeared at the April 2021 Shanghai Auto Show.

INTO EUROPE: THE ELECTRIC BRIDGEHEAD

For SAIC, the principal role of MG was to serve as its export automotive brand, sales initially building in new markets such as Eastern Europe, Africa, and Latin America—and, as we have seen, in MG's traditional U.K. home market. What this left out of the equation was the major market of continental Europe. SAIC took the strategic decision to launch MG in Europe as an electric brand, eschewing the usual petrol or diesel model ranges usually marketed by big names in that market. Starting in 2019, MG Europe began its journey with the MG ZS EV, MG HS PHEV, MG5 EV Estate (also sold in the U.K.), and, at the top of the range, an MG-badged version of what had already been marketed in China as the Roewe Marvel-R, a good-looking, quite upmarket electric crossover.

How MG Sales Have Mushroomed

Chinese MG sales figures for 2007 were just over three thousand. The following year they had tripled to over nine thousand, in 2009 to nearly fourteen thousand. Subsequent years have seen sales exceed twenty-nine thousand (2010), fifty thousand (2011), and so on—even allowing for the braking effect of the COVID crisis. By 2020, Chinese market MG sales were just less than three-hundred thousand, with 1.5 percent of the domestic market. There are no signs of the brakes coming off this relentless growth.

In the U.K., which some would still like to regard with nostalgia as MG's home market—where at least brand recognition remains strongest—MG stunned the market by selling more new vehicles in the first five months of 2022 than it sold in the whole of 2021. In the first months of 2023, the MG HS took top slot in the U.K. car market. Records keep falling—but nowadays they are sales rather than speeds.

TOP: In 2019, SAIC opened another Chinese factory, this time in Ningde, a port city on the northeastern coast of Fujian Province. The new facility was aimed specifically at exporting MG EV and PHEV production. Ningde Sanyu Park is only 2 miles (3.5 kilometers) from CATL, the world's leading EV battery supplier. SAIC already assembled MGs at factories in Lingang (Shanghai), Pukou (Nanjing), and Zhengzhou. *Author archive*

BELOW: The MG HS face-lift, now on sale in many markets, was previewed by this MG HS Trophy in 2020. Note the small grille badge placed on the big, bold grille shape. This is the next stage on from the Pekoe preceding generation's more classical chrome-framed grille. The ever-evolving modern MG range is increasingly international in focus, and sales in individual markets are frequently in the local top ten. *Author archive*

Unique MGs for Thailand and India

We saw earlier in this chapter how, in some markets, MG badges have been applied to Roewe cars. The only time this has been done in what might be termed a mature market was in the MG Marvel-X for Europe.

TOP LEFT: In some markets, including China, Australia, and—as with this example—in Thailand, this midrange MG five-door sedan continues to be sold as the MG5, while in other markets it is known as the MG GT. It was not sold in Europe, as it is not built in EV or PHEV form. *Author archive*

TOP RIGHT: What is known in India as the MG Astor, introduced there in August 2021, is basically the MG ZS in most other markets. *Author archive*

LEFT: The second-generation MG Extender, a pickup truck built and sold in Thailand. Essentially the same vehicle is marketed in other regions as a Maxus (another SAIC marque). *Author archive*

BELOW LEFT: After the MG Hector came the MG Gloster, another old aviation name. The basis for this large SUV was a Chinese vehicle from SAIC's Baojun brand, but it has been developed and restyled for the Indian market. *Author archive*

BELOW: A new addition to the Indian MG family in 2023 was the Comet, a tiny EV. *Author archive*

A face-lifted MG ZS EV appeared in 2021, offering principally a restyled nose and a larger battery. It arrived as sales were climbing rapidly worldwide, and this model was scoring top ten sales slots in many European nations. In the U.K., the MG5 EV Estate—which was face-lifted in 2022—proved a popular, practical family wagon, with a good combination of carrying capacity and a decent range. For MG in China, there were other new models built on a platform developed by SAIC, including the MG One, available in two parallel ranges with subtly differentiated styling. This model, however, has not been exported.

An important new model that *has* been introduced into Europe is the MG4, a stylish five-door EV hatchback that has earned both unprecedented critical praise and, in December 2022, a much-coveted safety affirmation through a five-star score in the independent European New Car Assessment Programme (NCAP) test. The MG4 was initially sold in China as the MG Mulan, named for a Chinese legendary heroine, and there was talk of a sporty four-wheel-drive MG Mulan Triumph variant said to offer a sub-four-

ABOVE: The MG XPower MG5 race car is the latest iteration of the racing MG sedan. *Author archive*

BELOW: MG's first completely new sports car in nearly thirty years, the all-electric MG Cyberster, made its public debut in roadgoing form at 2023's Goodwood Festival of Speed. The Cyberster would be on sale a year later; MG may yet offer further variants, and perhaps even a smaller EV sports car to sell alongside the Cyberster. *Author archive*

MG's "Safety *Very* Fast" offering, the MG4 XPower, launched in Europe in 2023, offers Porsche-rivaling acceleration, such as a 0–62 miles per hour (100 kilometers per hour) time of 3.8 seconds, delivered by engines front and back driving through all four wheels. *Author archive*

seconds 0–62-mile-per-hour (0–100-kilometer-per-hour) acceleration. The rate of growth of MG sales seems exponential, and, in Europe alone, sales rocketed by well over 100 percent in 2022 over the previous twelve months to achieve one-hundred thousand sales.

In two particular markets, however, MG has taken a slightly different path. In Thailand, SAIC's 50/50 partnership with local company Charoen Pokphand Group Company, SAIC Motor-CP, has spawned unique models such as the MG ExTender pickup (essentially an MG badged version of the Maxus T70 model sold in other markets), the MG5 GT hatchback, the MG EP (identical to Europe's MG5 EV Estate), and a model close to the ZS, badged the MG VS HEV. The MG V80 minibus is a rebadged Maxus V80 exclusive to Thailand.

Another particular market is India, where SAIC has established local production in a former General Motors plant. Here it builds a series of models derived from another SAIC-owned brand, Baojun, with a unique slant on turning these models into MG vehicles exclusively for Indian sales. These models have largely tapped into a perceived nostalgia for old U.K. aircraft names. Hence we see the MG Hector (launched in 2019), the larger MG Hector-Plus (2020), the MG Gloster (2020), and the MG ZS rebadged as the MG Astor for India (the all-electric version of the latter retains the original MG ZS EV name used in most other global markets). Anticipated in the near future is a much smaller entry model, a two-seater EV hatchback to be called the MG Comet.

CENTENARY AND BEYOND

If there is one thing that this concluding story tells us, it is that MG is not only back in a big way, but there seems little sign of the progress stalling any time soon. In July 2022, SAIC proclaimed its one-millionth export MG. If one adds this figure to previous milestones, stretching right back to 1923, then the number of MGs built by the time that the marque gained its century was in excess of 4.5 million.

Perhaps the most exciting aspect of all this busy activity for any longtime watcher of the MG brand is the promise of a return to MG's roots, presaged by that reveal in Shanghai of a showstopping electric sports car concept, the dramatic Cyberster. This innovative vehicle is aimed at a modern young audience, particularly the affluent Chinese middle class; many are tuned into today's big computer gaming lifestyle, as big in China as anywhere. The exterior of the Cyberster was largely the work of that London studio. The interior and much of the show tech was down to their colleagues in Anting, a suburb of Shanghai, which is home to vast design and production facilities, one of many such outposts across China.

A production version of the Cyberster, with powerful twin-motor four-wheel drive and 0–62-mile-per-hour (0–100-kilometer-per-hour) times well below four seconds, was previewed with full-sized display models in 2023, with various public outings including the MG Car Club's big annual jamboree at Silverstone in June that year. Definitive production versions followed not long after as part of the MG centenary celebrations.

The only question remaining is: When will SAIC return to sales in the United States and Canada? A new sales post has been established next door in Mexico. Surely it cannot be long before the United States is invited to rediscover "the sports car it loved first"?

Unveiled at the 2023 Goodwood Festival of Speed, the MG EX4, based on the MG4 XPower, was conceived as a tribute to the MG 6R4 and, in name the MG EX-E (see chapter 7 for both) of nearly thirty years earlier. *Author archive*

Acknowledgments

Every book depends upon not only the work involved in researching and writing it, but also the wider effort and support involved in turning it from a loose set of ideas to the finished article before you. This volume is no exception, and I would like to pay tribute to the many wonderful people who have contributed toward the outcome. Space precludes naming every individual, simply because in embracing over a hundred years of history, and writing about it for forty of those, there have been so many people along the way who have added to my knowledge and archives.

For the present volume, I would like to single out a few to thank. In terms of images, I am especially grateful to Christopher Keevill of the Early MG Society and to James Mann, automotive photographer par excellence, for their image contributions, which have greatly enhanced the visual aspects of this book. Christopher was also generous to proofread portions of the draft manuscript covering the beginnings of MG as did Peter Seymour, an expert and accomplished author on the topics of pre-war Morris, MG, and related matters—their shared wisdom and unstinting support is much appreciated. Over the years I have come to know and rely upon the friendship, support, and inside knowledge of a great many people who have been involved in the MG story at some point. Many who played a part in MG's earlier days are, sadly, no longer around to thank personally, but their contributions will always be remembered and appreciated by me. In some cases, I have later drawn support from their families—proud sons, daughters, and other relatives who have helped me try to do justice to their antecedents' stories. There are still a handful of individuals who bridge the gap between the old days of MG's Abingdon factory and the present day, and I am very appreciative of their enduring support: Peter Neal (MG Car Club archivist), Jim Cox (former MG worker and BMC Competitions insider), and Geoffrey Iley are amongst the surviving old guard, and I salute them. Through the Rover Group and MG Rover years, people who have helped me include Ian Elliott, Denis Chick, and Kevin Jones, just three of many who have supported my never-ending thirst for "modern" MG information.

Those involved with the present-day MG are happily still very much with us, and with that in mind I extend my thanks and gratitude to Guy Pigounakis and David Allison of MG Motor, as well as Carl Gotham and Rob Lemmens of the highly impressive SAIC Advanced Design London studio, as well as their talented and enthusiastic colleagues in China, notably Design president Shao Jing Feng, who has done much to drive the resurgence of the marque's sporting image.

Last but not least, all books rely greatly on the unsung heroes of the team involved in the editing, design, and publication, and I would like to bring Quarto/Motorbooks' Zack Miller and Brooke Pelletier and copy editor Tom Lewis into the limelight and say thank you for their skill and above all patience. I am proud of the outcome of all this collective effort, and I hope that you will enjoy it equally.

—David Knowles, Ruislip, U.K., 2023

Index

Quarto.com

First Published in 2024 by Motorbooks, an imprint of The Quarto Group,
100 Cummings Center, Suite 265-D, Beverly, MA 01915, USA.
T (978) 282-9590 F (978) 283-2742

Motorbooks titles are also available at discount for retail, wholesale, promotional, and bulk purchase. For details, contact the Special Sales Manager by email at specialsales@quarto.com or by mail at The Quarto Group, Attn: Special Sales Manager, 100 Cummings Center, Suite 265-D, Beverly, MA 01915, USA.

28 27 26 25 24 1 2 3 4 5

ISBN: 978-0-7603-8315-5

Digital edition published in 2024
eISBN: 978-0-7603-8316-2

Library of Congress Cataloging-in-Publication Data

Names: Knowles, David A., author.
Title: MG century : 100 years—safety fast! / David Knowles.
Description: Beverly, MA : Motorbooks, an imprint of the Quarto Group, 2024. | Includes index. | Summary: "MG Century, authored by marque expert David Knowles, offers a complete and richly illustrated history covering the evolution of this storied British car company"-- Provided by publisher.
Identifiers: LCCN 2023037392 | ISBN 9780760383155 | ISBN 9780760383162 (ebook)
Subjects: LCSH: M.G. automobiles. | M.G. automobiles--History.
Classification: LCC TL215.M2 .K56693 2024 | DDC 629.22209--dc23/eng/20230826
LC record available at https://lccn.loc.gov/2023037392

Design and Page Layout: Silverglass
Front cover: James Mann
Back cover: Jonathan Stein (upper image); SAIC Design (lower image)
Half title page: Martin Williamson
Full title page: SAIC Design
Table of Contents: Dimitri Dominguez
Endpapers: SAIC Design

Printed in China

ABOUT THE AUTHOR

David Knowles has become synonymous with MG history. He has a particular interest in the postwar story, when the marque exploded onto the world stage, becoming the preeminent sports car marque in North America, leading to the memorable advertising slogan: "MG. The Sports Car America Loved First."

By profession David is a Chartered Civil Engineer, but he has been writing about MG and Triumph sports cars (among other brands) for more than thirty years. Always turning up fresh information for his fellow enthusiasts, his previous works have covered every period of MG history, every model sold, and even cars only developed in prototype form. David's thorough research has expanded into the parallel field of postwar Triumph sports cars, including seminal works on the Triumph TR6 and the Triumph TR7 and TR8.

David has published fourteen MG books to date, several of which have been selected as "book of the month" by leading U.K. classic car magazines including *Classic Cars* and *Classic & Sportscar*. He is a Consultant Editor for *MG Enthusiast Magazine*. He has also contributed extensively to other internationally recognized MG publications. David resides in the U.K.

4